MOLECULAR
BIOLOGY
INTELLIGENCE
UNIT 14

Quasispecies and RNA Virus Evolution:
Principles and Consequences

Esteban Domingo
Centro de Biologia Molecular Severo Ochoa (CSIC-UAM)
Universidad Autonoma de Madrid
Cantoblanco, Madrid, Spain

Christof K. Biebricher
Max-Planck-Institute for Biophysical Chemistry
Gottingen, Germany

Manfred Eigen
Max-Planck-Institute for Biophysical Chemistry
Gottingen, Germany

John J. Holland
University of California, San Diego
La Jolla, California, USA

LANDES BIOSCIENCE
GEORGETOWN, TEXAS
U.S.A.

EUREKAH.COM
AUSTIN, TEXAS
U.S.A.

QUASISPECIES AND RNA VIRUS EVOLUTION: PRINCIPLES AND CONSEQUENCES

Molecular Biology Intelligence Unit

Eurekah.com
Landes Bioscience

Copyright ©2001 Eurekah.com
All rights reserved.
No part of this book may be reproduced or transmitted in any form or by any means, electronic or mechanical, including photocopy, recording, or any information storage and retrieval system, without permission in writing from the publisher.
Printed in the U.S.A.

Please address all inquiries to the Publishers:
Eurekah.com / Landes Bioscience, 810 South Church Street
Georgetown, Texas, U.S.A. 78626
Phone: 512/ 863 7762; FAX: 512/ 863 0081
www.Eurekah.com
www.landesbioscience.com

ISBN: 1-58706-010-8 (hard cover version)
ISBN: 1-58706-077-9 (soft cover version)

Library of Congress Cataloging-in-Publication Data

Quasispecies and RNA virus evolution: principles and consequences / Esteban Domingo, Christof K. Biebricher, Eigen, Manfred, John J. Holland.
   p. ; cm. -- (Molecular biology intelligence unit)
   Includes bibliographical references and index.
   ISBN 1-58706-010-8 (alk. paper)
   1. RNA viruses. 2. Viruses -- Evolution. 3. Viral genetics. I. Domingo, Esteban. II. Biebricher, Christof K., III. Holland, John J., IV. Series.
   [DNLM: 1. RNA Viruses -- genetic. 2. RNA Viruses -- pathogenicity. 3. Evolution, Molecular. 4. RNA Virus Infections -- genetics. QW 168 Q1 2001]
   QR395.Q376 2001
   579.2'5--dc21
                                                                    99-056721

CONTENTS

1. **Introduction** 1
   Darwinian Evolution as the Principle of Biological Dynamics 2
   Genotype and Phenotype 2
   Viruses and Other Genetic Parasites 3
   The Central Dogma of Molecular Biology 3
   Molecular Programs 4
   RNA and the Origin of Life 4

2. **Multiplication Strategies of RNA Genetic Elements** 7
   Biological Cloning of Viruses 9
   Four Basic Strategies of Virus Replication 10
   Strategies of RNA Virus Expression 11
   Prokaryotic RNA Viruses: Leviviridae 13
   Positive Strand RNA Viruses 15
   Negative Strand RNA Viruses 16
   Multipartite (Segmented Genome) RNA Viruses 18
   Subviral Agents 19
   Defective Viral Genomes 20
   Retroelements and Retroviruses 20
   Hepadnaviruses 22
   Disparate Life Cycles, Common Survival Strategies 23
   Early Evidence of RNA Virus Variability 24

3. **Molecular Recognition and Replication Enzymes** 28
   Structure of Nucleic Acids and Restrictions to Variation 28
   Molecular Recognition of Nucleic Acids, and Selection
     for Nucleic Acid Structures: Compensatory Mutations. 33
   Structure and Catalytic Properties of Proteins 34
   Quaternary Structures 37
   Replication Enzymes 38
   DNA-dependent DNA Polymerases 39
   RNA Replication 41
   Reverse Transcriptase 41
   Fidelity of RNA and DNA Polymerases
     and the Survival of Defective Genomes 42

4. **Quantitative Molecular Evolution** 47
   Selection of the Fittest 50
   Mutant Spectra and Sequence Space 51
   Sequence Heterogeneity 55
   The Sequence Space 56
   The Quasispecies 57
   The Error Threshold 60

5. **Darwinian Evolution of RNA in Vitro** 63
   The Kinetics of RNA Replication 65
   RNA Growth Rates 66
   Double Strand Formation 70
   Selection Among RNA Species 70
   Mutation in Replicating RNA 72
   Adaptation of RNA in Vitro 74
   Recombination Among RNA Molecules 76
   Can Biological Information Be Generated De Novo? 76
   Conditions for Replication 77

6. **Experimental Studies on Viral Quasispecies** 82
   Molecular Mechanisms of Genetic Variation of RNA Viruses:
      Mutation, Recombination and Genome Segment Reassortment 82
   Rate of Accumulation (or Fixation) of Mutations,
      and Its Lack of Regularity 89
   Population Equilibrium: Stasis in the Face of High Mutation Rates 93
   Population Disequilibrium: The Trigger of Rapid RNA
      Genome Evolution 94
   Unequal Rate of Occurrence of Different Mutation Types 96
   Hypomutation and Hypermutation 97
   Genetic Heterogeneity of Natural Populations of RNA Viruses 97
   Quasispecies Dynamics in Vivo 100
   Rapid Generation of Variant Genomes in Cell Culture 101
   Connection between Genotype and Phenotype. Viral Quasispecies
      as Reservoirs of Phenotypic Variants: Episodic Selection 102
   Thresholds for Phenotypic Expression 105
   Host Range Mutants of RNA Viruses 105
   Quasispecies Dynamics and Antigenic Variation 108

7. **Population Dynamics and Virus Adaptability** 119
   Repeated Population Bottlenecks Lead to Fitness Losses:
      Muller's Ratchet 120
   Molecular Basis of Fitness Decrease 121
   Fitness Increase of RNA Viruses: Quasispecies Optimization 123
   Multiple Molecular Pathways for Fitness Increase:
      a Wrightian View of RNA Virus Evolution 124
   Current Views on Fitness Evolution of RNA Viruses 125
   Complexity of Fitness Landscapes 125
   Evidence of Positive and Negative Selection Acting Continuously
      in Viral Populations 126
   Competition Between Neutral Variants: The Competitive Exclusion
      Principle and the Red Queen Hypothesis. Reproducible Nonlinear
      Population Dynamics 128

  Genetic Heterogeneity and Variability of DNA Viruses 129
  Extensions to Nonviral Systems: Mutation, Competition and
   Selection in Cell Populations 131
  Evolution of Bacteria. Resistance to Multiple Antibiotics as
   a Long-Term, Undesigned Experiment 132
  Evolutionary Potential of Unicellular Pathogens and Difficulties for the
   Control of Parasitic disease 133

8. **Connections, Implications and Prospects** 141
  Relationships of Quasispecies with Population Genetics and Current
   Concepts of Complexity 144
  RNA Viruses and Evolutionary Biotechnology 148
  Quasispecies and Viral Disease Control Strategies 158
  New Antiviral Strategies Based on Violation
   of the Error Threshold 158
  The Emergence and Reemergence of Viral Diseases 160
  Overview 162

 Index 172

PREFACE

Evolution is usually considered to be an historic process in which the living organisms found today have been shaped as a result of diversification from ancestor life forms in the time range of many millions of years. Most people, even many biologists, are not aware that evolution is a very important phenomenon in biology responsible for forming and altering populations of some types of cells and organisms in observable time spans. Evolutionary processes in viruses are at the root of their pathogenic potential. We have witnessed the emergence of as many as forty new pathogenic viruses in the last few decades, including the deadly human immunodeficiency virus (HIV) in our generation. As Charles Darwin showed, evolutionary adaptation to the environment takes place by selection of stochastically generated variant forms of existing individuals. When populations of an organism and its mutation rates are sufficiently large, the adaptation process is rapid enough to be observed in the field and in the laboratory, and to have important biological consequences. In these cases Darwinian evolution need not be inferred since it can be witnessed and even manipulated.

One of the most fascinating systems in which evolution and its consequences can be immediately felt are the RNA viruses. They cause important human diseases such as several forms of hepatitis, severe respiratory or hemorrhagic disease, cancer or AIDS. And in all these pathogenic processes viral genome evolution plays a role. RNA viruses by virtue of their high mutation rates and large population sizes form complex mutant distributions that have been termed viral quasispecies. The quasispecies concept was first developed by one of us (M.E.) to describe critical events in the generation and optimization of early life forms on earth. Investigation of quasispecies is a example of cross-fertilization between theoretical and experimental fields of research. Studies in the last three decades have been pursued at three levels: theory, molecular biology and virology. This book addresses connections between theoretical quasispecies and real virus quasispecies, and the biological implications of the quasispecies structure of RNA viruses and other simple replicons. It aims at reviewing evolutionary processes undergone by RNA viruses while replicating to produce disease. We will attempt to break the barrier that separates theoretical from practical studies with quasispecies, and to present the reader with updated information on basic evolutionary concepts and how they apply to pathogenic RNA viruses.

It is also increasingly evident that the key processes of rapid genetic variation, competition and selection which result in short-term adaptability are not restricted to RNA viruses. DNA viruses, bacteria, pathogenic cellular parasites, immune cells and cancer cells can be observed to exploit the same processes, but the time intervals in which the phenotypic alterations become evident are different. It has taken half a century for the selection of antibiotic-resistant bacteria to represent a widespread threat to humans, and yet it takes only weeks to months to select inhibitor-resistant immunodeficiency viruses in treated patients. Yet the basic underlying evolutionary phenomena are of the same nature.

It is the aim of this book to provide a broad picture of the principles common to rapid evolution, together with the problems stemming from the adaptability of pathogenic agents. The book has been divided into eight chapters covering a variety of topics related to molecular evolution and virology, unavoidably with considerable variation in depth. The reader will find several references at the end of each chapter that should help in finding additional relevant information. We have made an effort to stick to solid evidence, and to distinguish models from reality. We have also aimed at quoting general review articles and books, but occasionally very specific papers are cited to underline certain concepts with a detailed study.

The number of references given is necessarily limited, yet many colleagues have contributed to the theoretical and experimental studies summarized in the book. Our gratitude goes to all who have participated and are participating into this fascinating field of research, and who have provided useful information in the form of reprints or in discussions. We acknowledge the excellent assistance of Lucia Horrillo with the preparation of the manuscript.

Esteban Domingo
Christof K. Biebricher
Manfred Eigen
John J. Holland

CHAPTER 1

Introduction
The Ever-Changing Nature

It has always been the main aim of human intelligence to attempt to understand the bewildering diversity of the environment by recognising patterns and regularities of events. It is thus no accident that one of the first objects of scientific observation was the starred night sky: it displays to the eye a fantastic pattern of light points of different luminosity and color. A persevering observer can note that despite its apparent randomness, the pattern of light points does not change from observation to observation: the relative orientation of the different points apparently remains the same. An immediate sense of the pattern is not recognizable, and the earliest astronomers tried to bring some order into the pattern by constructing constellations modelled after simple objects.

Despite the apparent invariance of the geometrical orientation of the stars, there is a highly regular periodic change recognizable: exact observations revealed that the pattern rotates during the night and that the observed pattern is exactly reproduced after a period of one year, suggesting the division of the whole cycle into 360 degrees. Even for the planets that do not follow the simple mechanical rules that apply to distant "fixed" stars, a suitable model that was in full agreement with the observations was presented by Ptolemaeus. However, later measurements with higher accuracy revealed phenomena that could not be reconciled with the model of Ptolemaeus. The scientific concepts of astronomy had to be revolutionized several times, and the change of paradigm did not occur without controversy and serious disputes. Today we know that the apparent invariance of the pattern is fortuitous: there is neither direct interaction between the stars of a constellation nor are the patterns invariant: because of the enormous distance to the stars the change in the pattern is so slow that it escapes observation unless it can be demonstrated by measurements of very high precision. The static picture of astronomy was slowly tranformed into a dynamic concept where stars and even galaxies are formed and die, culminating in the startling realization that even the universe had an origin and may one day have an end. In a similar manner, modern plane tectonics has transformed our view of the crust of our own planet, earth, from a rather static structure to a rather dynamic, ever-changing environment.

The notion of an ever-changing nature has influenced human thinking as reflected in philosophy and science in practically every discipline. In biology, the concept of the origin of the species by evolution, first formulated by Charles Darwin, was so appealing and successful in interpreting biological phenomena that many years later Dobzhansky made his famous statement "Nothing in biology makes sense except in the light of evolution" (Dobzhansky, 1973). We live in a continuously evolving, complex and to a large extent unpredictable universe. Our constantly-evolving environment necessarily demands that life forms must also evolve (or perish).

Quasispecies and RNA Virus Evolution: Principles and Consequences, by Esteban Domingo, Christof K. Biebricher, Manfred Eigen and John J. Holland. ©2001 Eurekah.com.

Darwinian Evolution as the Principle of Biological Dynamics

Most historians will agree that the cradle and the basis of human civilization were the developments of techniques for stock-farming and agriculture, requiring the (artificial) selection of plants and animals that had suitable properties for production and for nourishment (see, for example, accounts in Maisels, 1993 and Fagan, 1996). It was recognized that concomitant with the increased utility of a species was a loss in the animal's of ability to survive under natural conditions, and that a domesticated species left to competition in nature had to return to its "wild type" traits or perish.

Despite the impressive technical knowledge for selection of cultivars and animal strains, it took millenia to realize that species in nature also undergo a rigorous selection and that this selection is the driving force of evolution (Darwin, 1859). However, it was rather the other, more passive force in evolution, mutation, that was not understood. The unforced and undirected generation of mutants was the revolutionary idea of Darwin; it contradicted the general understanding of nature by violating the human feeling that everything has an aim or direction.

Genotype and Phenotype

While the ideas of Darwin have been enormously successful in biology, reports about shortcomings of Darwin are frequent, and are often exaggerated in nonscientific newspapers. There can be no doubt that some more detailed statements of Darwin are not in agreement with what we know today. That is no surprise: at the time Darwin published the *Origin of Species* our understanding of genetics was rudimentary. The simplest laws of heredity were found at about the same time by Gregor Mendel, but the impact his ideas had on his contemporary biologists was negligible. Only a generation later were serious genetic studies resumed, and these led to the understanding of some underlying basic laws of evolution. This synthesis and melding of genetics and evolution was advanced by schools such as those of Haldane, Fisher, Wright and others (reviewed in Dobzhansky *et al*, 1977).

For a long time it was common agreement that the transformation of species proceeds over huge time spans that are not suitable for direct observation. The investigation of viruses, in particular of RNA viruses, has changed this view. RNA viruses not only adapt rapidly to changes in environment (Holland *et al*, 1982, 1992); we have witnessed in the last decades the emergence of new RNA virus species, often with dramatic consequences due to novel pathogenic properties (Morse, 1993). Ultimately, scientists had to recognize that the very adaptability of RNA viruses made the fight against the new diseases they cause exceedingly difficult. Antiviral drugs which apparently offered successful treatment often proved worthless after a short time due to the emergence of drug-resistant viruses, and pharmacologists soon wished that viral evolutionary adaptation would require thousands of years! While the basic evolutionary principles are the same for all biological forms, the pace at which evolution progresses, and the time spans required for major evolutionary events to occur are exceedingly different for organisms of the classical biological kingdoms as compared to the viruses, and we shall discuss the reasons for this in the following Chapters.

In evolution, the genotype of an organism, characterized by the genetic material contained in its genome, should be distinguished from the phenotype, whereby the genome manifests itself in response to its environment.

For a long time, genetics meant Mendelian genetics, where sexual crosses and observation of phenotypic markers were the main methods to approach evolution. The advent of molecular biology in the second half of the twentieth century presented the possibility for more precise measurements. Today, the genotype is clearly defined as the sequence of nucleotides in the genome, which can be precisely determined with reasonable experimental efforts. The already quite large sequence library has been exponentially amplifying in the last few years. Less impressive is our present knowledge concerning the expression of the genotype in the phenotype. The enormous complexity of the biochemical processes going on, and the complications introduced by differentiation and development of multicellular organisms allow, at best, a very crude assessment of the phenotypic properties of a genotype. We refer to the complex reactions that are programmed by the genome as expression. It requires an enormous apparatus for decoding the digital information of the genome, and thus the smallest unit of life is the genome together with its decoding machinery, i.e., the cell.

Instrumental in the earliest studies of expression of genotypes were biological objects or (depending on the definition) organisms that have small genomes and thus only relatively few functions which can be expressed. They have been widely characterized, starting with the first quantitative experiments of Delbrück and colleagues with bacteriophages (reviewed in Cairns et al, 1992). Virus genomes are small, their population size is often high, and their mutability is particularly high. Therefore, viruses were not only instrumental in the study of molecular genetics, but also in the study of evolution. While the origin of new species in the animal or plant kingdoms requires enormous time spans, several "new" virus species have emerged in our life time.

Viruses and Other Genetic Parasites

What makes the viruses so unique? It is not only the size of their genomes, which are much smaller than those of the most primitive cellular organisms; it is also their genetic organization. Viruses are intracellular parasites and multiply in the cell as do genetic elements of the host. The "classical" virus has a defined, often regularly shaped infectious particle, the virion, containing an RNA or DNA genome surrounded by a protein shell, sometimes enveloped in a membrane. In this state, a virus is unable to replicate. Its only reactions are the recognition of a receptor on a suitable host cell, triggering the invasion of the virus itself (or at least its genome) into the host cell. Once inside the cell, the viral genome uses the genetic apparatus of the host to decode the viral information, to amplify the genome and its encoded products, and to assemble the viral components to mature virions, which eventually are liberated from the host to start a new infection cycle.

The organisation of viruses is enormously flexible (Chapter 2). Sometimes, the viral genome becomes stably integrated into the host genome and may be dormant for many cell generations, until it suddenly enters a "productive" growth phase. Parts of the viral genome can be lost and the "cryptic" virus may survive for prolonged time periods as an autonomous genetic element in the host without ever producing a virion. The most primitive virus-like molecules known are the plant-pathogenic viroids, naked RNA molecules of rather short chain lengths, only 300-500 nucleotides long.

The Central Dogma of Molecular Biology

Genetic information is digitally stored as a sequence of nucleotides, similar to a computer program. In order for a computer program to be executed, it needs the appropriate hardware and an operating system that decodes and interprets the information and provides the basic commands and operations. It is a characteristic of life that the genetic program includes not only the full operating system but also the prescription to build hardware from certain components which must be provided by the environment. For execution, however, both the hardware and the operating system must be present; for this reason, cells are sometimes regarded as the "atoms of life".

If this feature is considered as a necessary condition for life, then viruses are not living because they are dependent on the hardware and the operating system of the host. Only the appropriate genetic apparatus can execute the genetic program of the virus, i.e., viruses are strongly host specific and host-dependent.

The fundamental processes of the operating system of a cell are combined to the so-called "central dogma" of molecular genetics: The information is stably stored in a double-stranded

DNA genome and copied by a process called *DNA replication*. That part of the information that is needed at a given time to fulfill a certain task is copied in a process called *transcription* into working copies of the information, the messenger RNA (mRNA). The nucleotide sequence is *translated* into a sequence of amino acids, according to the rules of the genetic code. The resulting proteins serve as highly specific catalysts or structural components for the metabolic and other chemical and physical reactions of the cell.

While all cellular organisms contain DNA as genetic information, a large percentage of known viruses have RNA genomes, usually as single strands. RNA genomes must be amplified for viral spread, but RNA amplification is not among the processes of the central dogma. The virus has thus to provide for its own replication system. There are several strategies: In the "plus strand viruses", the viral genome is used directly as mRNA and the viral RNA is replicated by a virus-encoded RNA "replicase". In the "minus strand viruses", it is transcribed into different mRNA molecules and a full length plus strand which is subsequently used as template to produce minus strands to be packed into virions. In the retroviruses, the viral RNA is transcribed by a viral "reverse transcriptase" into a double-stranded DNA, which may be integrated into the host genome. From the DNA, different mRNA molecules and full-length plus strands can be made; the latter are packed into the virion shell (Chapter 2).

Why are RNA viruses so successful in evolution, while not a single cellular organism with an RNA genome has survived? We shall see later that the answer lies in the limited genome size and the high mutability which provides for rapid adaptation of viruses to their environment. High mutability would not allow the stable conservation of the large information content that is required for the complicated network of biochemical processes and structural components in a cell (Chapter 4).

Molecular Programs

The nucleotide sequence of an organism is a genetic program: Biochemical reactions are started that trigger other reactions forming cascades of consecutive expression steps. The expression of the genetic program thus changes with time, defining an "age" of the organism correlated with the developmental status of the program. The more genes an organism has, the more complicated is its program. Since viruses have only a few genes, the cascade of chemical steps are relatively easy to investigate, and the steps constituting the "molecular program" of many viruses, in particular of bacteriophages, have been identified. Usually the program starts with the infection of a host cell by a virus particle and ends with the release of the progeny virions (Chapter 2).

RNA and the Origin of Life

Self-reproduction, the basis of evolutionary adaptation, serves two purposes: First, it allows conservation of information even though its molecular carrier is subject to chemical modification. Second, its autocatalytic nature provides an efficient selection mechanism. Only nucleic acids possess the inherent chemical potential for self-production and there are strong indications that the first information carriers in early evolution were RNA (or RNA-like) rather than DNA molecules (Joyce and Orgel, 1993):

- Ribose is readily formed under potential prebiotic conditions by aldol condensation of formaldehyde.
- The vicinal 2',3'-hydroxyl groups of ribose are stronger nucleophiles than the 3'-OH of deoxyribose, resulting in a higher efficiency of phosphodiester formation between ribonucleotides.
- The metabolic biosynthesis of deoxyribonucleotides proceeds via reduction of ribonucleoside diphosphates. DNA polymerases are unable to initiate nucleotide condensation; DNA replication therefore also requires RNA synthesis (Chapter 3).

- Ribonucleotide base pairs are structurally more stable than their deoxyribo analogues. RNA single strands usually have a strong secondary structure which is further stabilised by tertiary interactions, often involving hydrogen bonds with the 2'-hydroxyl of ribose. The higher energy of tertiary interactions favours various single stranded forms of RNA, while DNA prefers fully base-paired double strands.
- The single-stranded forms of RNA offer a rich repertoire of tertiary structures, allowing surfaces with highly specific binding and specific catalysis. All functional nucleic acids involved in protein biosynthesis and RNA processing are RNA molecules. Specific catalysts consisting of RNA have been well characterized. They resemble protein enzymes and are thus named "ribozymes". A large variety of additional catalysts have been selected by evolutionary biotechnology (Chapter 8). Some scientists have referred to an ancient "RNA world", where RNA catalyzed all basic reactions in life, perhaps aided by nonspecific cofactors, to be a necessary intermediate in the origin of the progenote cell. Later, the invention of the translation process allowed the synthesis of more efficient catalysts. The functional RNAs in the cell may be remnants of this RNA world.

There are several reasons why DNA replaced RNA as information carrier at a later stage of evolution:
- RNA is chemically less stable under ambient aqueous conditions due to its moderate rate of hydrolysis catalyzed by organic and inorganic bases and divalent metal ions.
- DNA allows error correction by using the complementary strands for reference.
- The predominant chemical reaction modifying the nucleobases is deamination of cytidine to uridine. In DNA, uridine is replaced by thymidine in order to detect and repair these changes.
- DNA and RNA fold into different double-stranded structures. It is easier to open the DNA structure and to recognise base pairs in the double helix.

It is thus plausible that DNA replaced RNA when the conservation of large amounts of information was required. On the other hand, present-day RNA viruses may not be considered to be remnants of the RNA world, because they require the preexistence—or at least the concomitant evolution—of their hosts. Nevertheless, the properties mentioned above apply to them and they have been shown to be excellent model systems for experimental studies of molecular evolution.

In the next Chapters evidence of the impressive evolutionary potential of RNA viruses and other RNA genetic elements is presented. It is impressive to the point of casting doubts on assertions that with our current knowledge of chemistry and biology there has been insufficient time in the earth history to explain the full development of complex life forms that we see today. This view may have to be modified once a more complete picture of how RNA genetic elements can interact with differentiated organisms is acquired. And more so considering that retroid agents have penetrated and agitated the cellular world since primordial cellular organizations provided the basis for complex life forms.

References

1. Cairns J, Stent GS, Watson JD, eds. Phage and the Origins of Molecular Biology. Expanded edition. Cold Spring Harbor Laboratory Press, 1992.
2. Darwin C. The origin of species. 1859. (Many editions) New York: Mentor Books, The New American Library, Inc. 1958.
3. Dobzhansky TH. Nothing in biology makes sense except in the light of evolution. Amer Bio Teacher 1973; 35:125-9.
4. Dobzhansky TH, Ayala FJ, Stebbins GL et al. Evolution. San Francisco: Freeman, 1977.
5. Fagan BM. World Prehistory. A Brief Introduction. New York: Harper Collins Publishers Inc., 1996.

6. Holland JJ, Spindler K, Horodyski F et al. Rapid evolution of RNA genomes. Science 1982; 215:1577-85.
7. Holland JJ, de la Torre JC, Steinhauer DA. RNA virus populations as quasispecies. Curr Top Microbiol Immunol 1992; 176:1-20.
8. Joyce GF, Orgel LE. Prospects for understanding the origin of the RNA world. In: Gesteland RF, Atkins JF, eds. The RNA World. Cold Spring Harbor Laboratory Press 1993:1-25.
9. Maisels CK. The Emergence of Civilization. London and New York: Routledge, 1993.
10. Morse SS, ed. Emerging Viruses. Oxford: Oxford University Press, 1993.

CHAPTER 2

Multiplication Strategies of RNA Genetic Elements
An Overview of the Viral Life Cycle

The main steps in the life cycle of a virus are (Fig. 2.1): receptor and coreceptor recognition at the cell surface, entry into the cell, uncoating and release of the genetic material, viral gene expression, viral genome replication, assembly of progeny particles, and exit from the cell with the acquisition of membrane material from the cell in the case of enveloped viruses.

A virus must recognize a cell receptor to interact with the cell and to enter the cell. Specific protein domains on the virus surface are endowed with the potential to recognize a cell receptor molecule. The process is an important determinant of virus tropism and therefore of host-cell specificity, as documented in early studies with bacterial viruses and poliovirus (McLaren et al, 1959; Holland, 1961; Crawford and Gesteland, 1964; review of early work in Longberg-Holm and Philipson, 1974). In the last decade, many studies have gradually modified the early views that every virus had a single specific, matching receptor, into the view that many viruses can use a number of different receptors and coreceptors to attach to cells albeit with different efficiency (Evans and Almond, 1998). Perhaps the most dramatic example is provided by the human immunodeficiency viruses (types 1 and 2) which, in addition to using the surface antigen CD4 as the main receptor, can utilize a broad range of chemokine and cytokine receptors (Alkahtib et al, 1996; McKnight et al, 1998; Sattentau, 1998). Viral receptors are frequently abundant cell surface molecules (proteins, carbohydrates or glycolipids) that perform basic functional roles for the cell (internalization of molecules, immune responses, etc.). MHC class I and class II proteins are used as receptors by some togaviruses, sialyloligosaccharides by influenza virus, intracellular adhesion molecule 1 (ICAM-1) by human rhinoviruses, CD4 by human immunodeficiency virus, heparan sulfate by some herpes viruses, α-dystroglycan by some arenaviruses, and integrins by foot-and-mouth disease virus, among many other examples (Wimmer, 1994; Mondor et al, 1998; Cao et al, 1998; Evans and Almond, 1998; Schneider-Schaulies, 2000). Expression on the cell surface of a protein known to be a receptor for a virus does not ensure that this cell is going to be infected by the virus. Sometimes additional surface molecules are required for the virus to enter the cell. The presence of the poliovirus receptor in different tissues does not always confer susceptibility to the virus (Evans and Almond, 1998), and there is evidence that additional molecules are required for infection. In general, functional receptors for a virus are expressed in many more types of tissues and cells than those that are actually infected by the virus. Following efficient penetration into the cell, an intracellular block may preclude progression of the infection. One of the receptor groups for foot-and-mouth disease viruses (FMDV) are the widely distributed integrins. Yet FMDV infects only artiodactyls, and its host range in cell culture is also limited. Amino acid substitutions at the

---

Quasispecies and RNA Vvirus Evolution: Principles and Consequences, by Esteban Domingo, Christof K. Biebricher, Manfred Eigen and John J. Holland. ©2001 Eurekah.com.

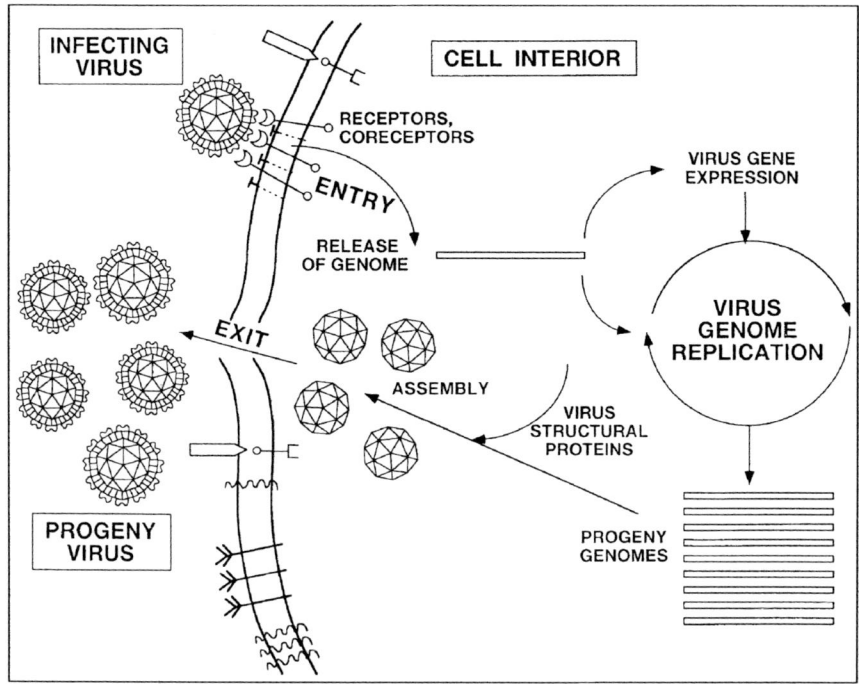

Fig. 2.1. A schematic overview of a virus life cycle. Essential steps are receptor recognition and entry into the cell, viral gene expression with the net result of production of many progeny viral genomes, assembly of genomes into progeny particles and exit from the cell. The steps involved in the life cycle of viruses are discussed in the text. Information on most experimental fields of animal virology can be found in Fields et al (1996) and Flint et al (2000).

receptor-recognition domains of viruses can mediate changes in receptor specificity. These changes are believed to contribute to modifications in the host range of viruses and to the emergence of new viral pathogens (Chapter 6). Adaptation of FMDV and some other viruses to cells in culture often results in amino acid substitutions at the capsid surface that enhance the ability of the virus to interact with heparan sulfate and lead to a change in cell tropism.

Following receptor recognition, a virus enters the cell in an irreversible manner, through an energy-dependent process. Major mechanisms are endocytosis and virus-cell fusion. In endocytosis, an endocytic vesicle containing the virus is formed. Acidification of the internal part of the vesicle (endosome) initiates a chain of modifications which results in the release of the naked DNA or RNA genome or of the nucleocapsid (a complex between the nucleic acid genome of the virus and proteins) into the cell cytoplasm. Fusion of enveloped viruses with the cell is mediated by viral surface proteins (the "fusion" or F proteins, hemagglutinins or neuraminidases, or by the concerted action of several of these proteins). Fusion is prompted by conformational changes of the relevant proteins and the result is the release of the nucleocapsid of the virus into the cytoplasm. The process is linked to the uncoating of the viral capsid to liberate the viral genome. In some viruses, the entire capsid is transported into the cell nucleus where viral replication takes place. It is worth noting that virus capsids and envelopes must be sufficiently stable to protect the viral genome from the extracellular environment, but the particle must also be capable of disassembling upon entry into a host cell.

Genome replication varies substantially for the different viral groups, and the main strategies are briefly summarized in the following sections of this Chapter. As indicated in the scheme (Fig. 2.1.), in all cases, proteins encoded by the viral genome (be it a DNA or an RNA virus) must be expressed in order to achieve viral genome replication. Normally both viral and cellular proteins participate in virus multiplication. The cellular factors are generally involved in RNA processing or translation of cellular RNA, and they are captured for viral RNA replication (Kamen, 1975; Lai, 1998). Progeny genomes must be encapsidated to form progeny virus particles that are released from the infected cells, often resulting in cell death. Huge amplifications of viral genetic material are achieved in each acutely infected host cell. Some interactions of viruses with cells may lead to nonproductive infections in which the viral genome may be lost or integrated into cellular DNA (as in the case of retroviruses). As a result of a nonproductive infection, the cell may die or survive. Surviving cells may be altered in their growth properties, showing oncogenic transformation. Virus production may be limited in the case of persistent infections in which replicating viruses and dividing cells can survive for prolonged time periods. In persistently infected cell cultures, a coevolution of cells and the resident virus is sometimes seen, and the cells may evolve towards a transformed phenotype (loss of contact-inhibition of cell growth and ability to form cell colonies in semisolid agar or tumors in animals).

Productive viral infections are generally manifested by a cytopathic effect on cells. This often involves cell rounding, detachment from the solid surface, and cell lysis. In other cases there is formation of syncytia (multinuclear, fused cells) or inclusion bodies (aggregates of macromolecular structures in the cell interior). It is not easy to diagnose the reasons why cells die following viral infection. Cytopathic effects are believed to be due mostly to secondary effects of viral replication rather than to the toxic effect of a specific virus-coded product. There are two pathways for cell death, necrosis and apoptosis (or programmed cell death), and both can participate in causing the death of a virus-infected cell. Viral proteins able to modulate the apoptosis pathway have been described in viruses, and they may be important to facilitate persistent infections. For detailed reviews of the infectious cycles of RNA and DNA viruses the reader should consult Fields et al (1996) and Flint et al (2000). Additional references are given in the following sections.

Biological Cloning of Viruses

Early work by Ellis and Delbrück 60 years ago established that a single phage particle was sufficient to initiate a productive infection in *Escherichia coli*. This was shown by the linear increase of plaque counts with the relative concentration of the virus preparation. A plaque is a zone of cell lysis on a cell monolayer caused by the progeny of a single infectious virion. Dulbecco extended this methodology and allowed the quantitation of infectivity of animal viruses by plating serial dilutions on monolayers of animal cells (Fig. 2.2). Virus preparations may sometimes contain aggregated particles that could produce progeny from different parental genomes with the same plaque morphology that would be produced by a single particle. Virions may be disaggregated by treatment with mild detergent prior to dilution and plating. Some multipartite viruses, notably plant viruses, require coinfection by more than one particle for productive infection. A few important viruses, notably some human hepatitis viruses, have not been grown efficiently in cell culture. However most viruses do so, and can be subjected to biological cloning, irrespective of their replication strategy. Comparison of viruses from individual plaques, and studies on the evolution of viral populations derived from single infectious particles, have been instrumental in the development of the quasispecies concept. Also, important concepts in the dynamic behavior of viral quasispecies have been established using biological clones of viruses. Progeny of a virus from a single plaque have undergone the most severe form of genetic bottleneck: a transient reduction of the viral population size to one single infectious genome.

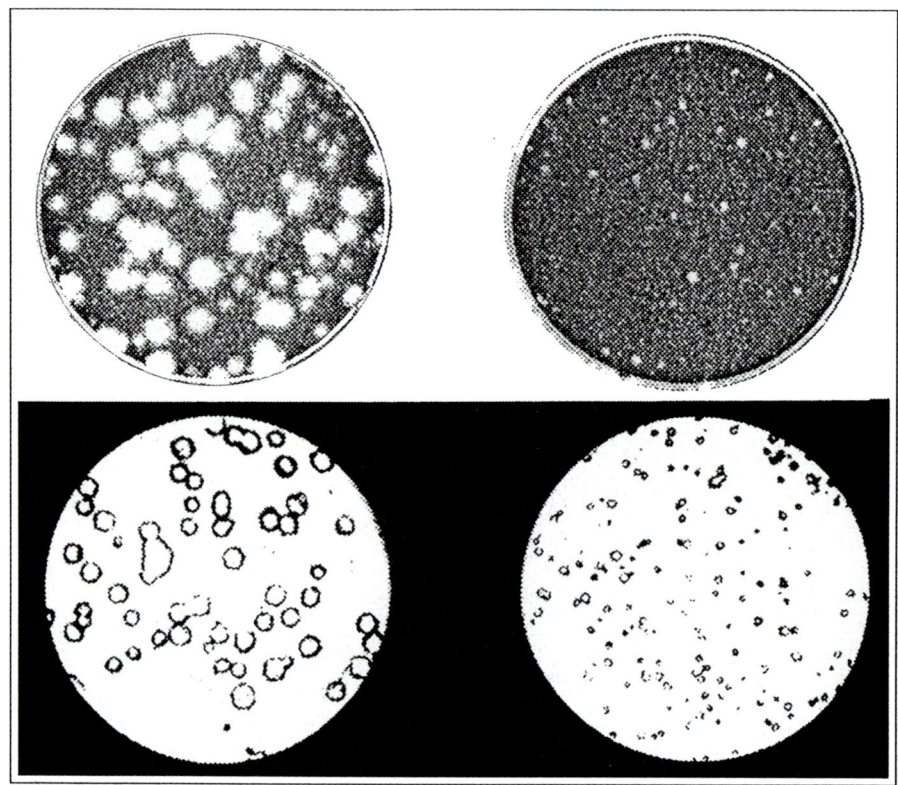

Fig. 2.2. FMDV plaques on BHK-21 cell monolayers visualized by two different procedures. Top: After 40 h of plaque development under an agar overlay, the cells were fixed and stained with crystal violet. Plaques are seen as regions of unstained (killed) cells as a result of virus infection. Bottom: Monolayers were transferred to a nitrocellulose filter and plaques visualized by ELISA with a FMDV-specific monoclonal antibody (details in Díez et al., 1989). Note the different plaque morphology (large, small plaques) of FMDV genetic variants. Plaques are frequently visible without staining and infectious virus can be isolated from them, constituting a biological cloning of the population (see text). (Bottom picture adapted from Díez et al., 1989, with permission from SGM, U.K.).

Four Basic Strategies of Virus Replication

Both DNA and RNA viruses use a complex array of biochemical reactions, the great majority of them inside the host cells, to ensure their survival as molecular parasites. This network of biochemical reactions is enormously complicated. Evolutionary change and adaptation to a diverse cellular world has shaped a variety of replication designs aimed at producing hundreds to hundreds of thousands of progeny viral particles from a single parental genome in an infected cell. Four general strategies for virus replication have been identified (Fig. 2.3). In strategies 1) and 3) of Fig. 2.3, the viral nucleic acid as well as all replication intermediates are of the same type, RNA or DNA. Strategy 1) is used by most prokaryotic RNA viruses and many animal and plant viruses, several of them causing important diseases. These include poliovirus, influenza virus, hepatitis A and C viruses, rotaviruses, yellow fever virus and the Ebola filovirus. Strategy 3) DNA replication is used by all living cells, and many DNA bacteriophages have adopted it; examples are the T-bacteriophages, lambda, and the animal herpesviruses, the poxviruses and the papillomaviruses.

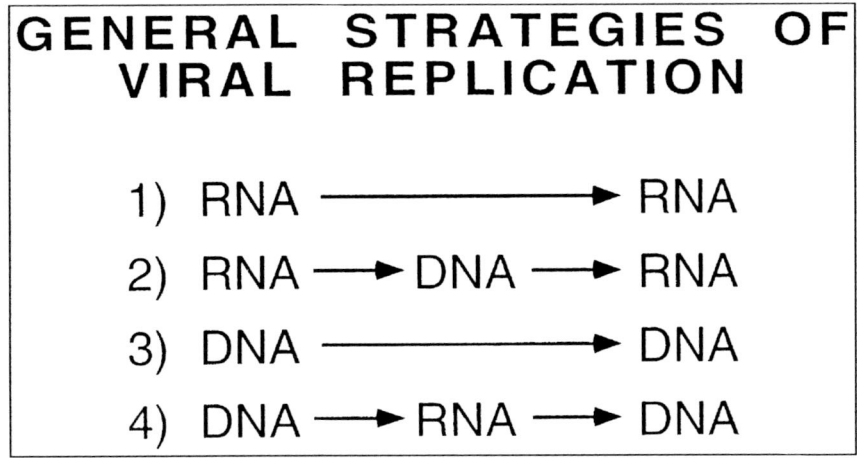

Fig. 2.3. The four basic strategies for viral genome replication. Arrows point to the flow of genetic information of the virus in the infected cell. The nucleic acid written in the first place is the one found in the virus particle.

In strategies 2) and 4) a type of nucleic acid different from that found in virions is synthesized as a replicative intermediate. Enzymes that synthesize RNA from DNA are called transcriptases; those that synthesize DNA from an RNA template are called reverse transcriptases. Strategy 2) operates in retroviruses, the human immunodeficiency viruses associated with AIDS being a dramatic, but by no means unique, example of a disease-causing retrovirus. Strategy 4) is used by the hepadnaviruses which include the important human pathogen hepatitis B virus and the plant virus cauliflower mosaic virus. In this case, a viral polymerase catalyses the synthesis of genomic DNA from a pregenomic RNA intermediate.

DNA viruses contain usually double-stranded genomes while RNA viruses prefer single-stranded genomes, but single-stranded DNA viruses and double-stranded RNA viruses occur too. Since base-pairing is the molecular principle underlying all replication strategies, both complementary strands must be involved in a full replication cycle. In single-stranded RNA genomes there is a clear distinction between the two complementary strands: usually, the one that serves as mRNA is named sense, plus or positive strand, while the other one is called antisense, minus or negative strand. In this Chapter we examine in more detail the replication strategies of RNA viruses, including DNA viruses which use RNA as a replication intermediate [strategies 1), 2), 4) in Fig. 2.3].

In addition to the fully competent RNA viruses, animals and plants are able to replicate, sometimes autonomously, sometimes with a helper virus, a variety of subviral RNA elements. Altogether, RNA genetic elements are the most abundant group of molecular parasites known. It is estimated that at least 75% of all pathogenic viruses are RNA viruses, and most emergent and reemergent viral diseases are associated with RNA viruses (Chapter 9). Let us examine replication strategies and some evolutionary implications of RNA genetic elements.

Strategies of RNA Virus Expression

Apart from the DNA viruses which use essentially the expression mode formulated in the central dogma (see previous Chapter), the simplest strategy for a virus is to act as a parasitic messenger RNA. Upon entering the cell, the mRNA expresses its proteins, replicates and is packed into mature virions which are liberated by the burst of the host cell. Since the messenger is

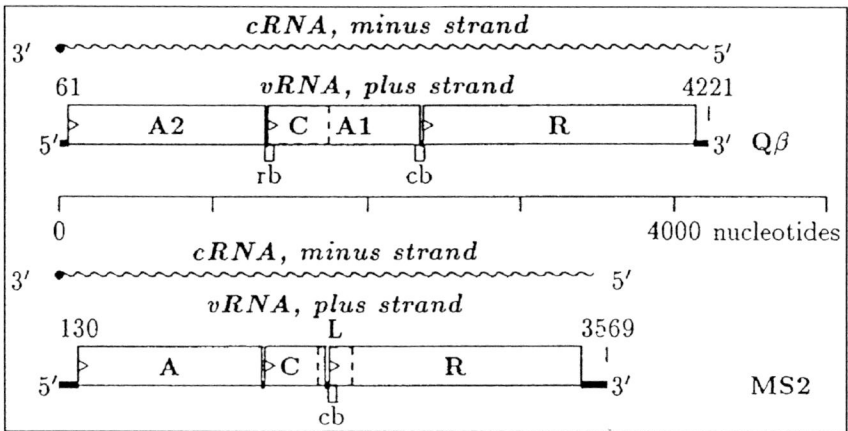

Fig. 2.4. Gene map of leviviruses Qb and MS2: ribosome binding site; ⟨: replication start site. rb: replicase binding site; cb: coat protein binding site. Phage MS2 probably has an analogous replicase binding site. Note that the specific protein binding sites and the ribosome binding sites are very close and even overlap in some places. A1-protein is made by occasionally reading through an *opal* stop codon. L protein is made by occasionally reading a tetraplet codon in the C-gene, causing a frame shift leading to abortion by a stop codon and an immediate synthesis of a new protein in the new codon frame.

packed in the virion, these viruses are called positive strand RNA viruses. Indeed, the smallest of the viruses in the prokaryotic, plant and animal kingdoms follow this seemingly simple mechanism. During the infection cycle, the RNA has to participate in three fundamental functions—protein synthesis, replication and packaging into virions—and conflicts between these processes must be avoided. Furthermore, the different gene products are required in highly different amounts and at different times. Structural (S) proteins that are involved in building up the viral shell are required in at least stoichiometric amounts to the viral RNA, while nonstructural (NS) proteins catalysing synthetic steps usually should be present in smaller quantities. Furthermore, most of the enzymes involved in replication are required early in the infection cycle while packaging to form mature progeny particles is a late step in infection. The genome carries thus not only information for gene products, but has gene loci responsible for the regulation of processes, e.g., ribosome binding sites or sites where the coat protein binds to build up the mature virion shell. Figure 2.4 illustrates gene maps and critical regulatory elements for bacteriophages Qβ and MS2. While the gene products are freely diffusible in the host compartment and defects in one genome thus can be complemented in *trans* by another functional gene product, defects in the genome are manifested in *cis*, i.e. they affect the strand carrying the defect and can not be complemented by another "helper" genome. The genes that are expressed into proteins are called cistrons.

For prokaryotic RNA viruses the mRNA is polycistronic, i.e., several proteins can be made from one and the same mRNA. Regulation of the frequency of translation and translational repressors are involved in regulated expression. Most eukaryotic viruses do not follow this strategy because eukaryotic mRNAs are often monocistronic, i.e., only one protein can be made from a mRNA. This limitation led to several quite diverse strategies. The apparently simplest solution was to divide the genome into different RNA segments (multipartite genomes). This strategy was particularly successful in the plant kingdom, probably because the transport of viral material from cell to cell in one and the same organism could make use of an alternative called mobilization that makes it unnecessary to assemble a mature virus particle to infect a new host cell. Among animal plus strand viruses, this strategy was not competitive.

A particularly successful solution was the synthesis of a large "polyprotein" which is further processed by proteolytic cleavage into the required structural and catalytic proteins. This strategy seems to result inevitably in the production of all proteins in equal amounts and at the same time. This disadvantage is partially compensated by additional features: (i) The structural proteins are arranged at the amino terminus of the "polyprotein" so that the inevitable premature translational stops lead to maximal amounts of the structural proteins and to reduced amounts of the nonstructural ones. (ii) Proteolytic processing involves a complex cascade of reactions, and catalytic proteins can be further degraded to inactive products leading to reduced steady state concentrations of catalytic proteins. (iii) For some viruses site-specific translation attenuation where the ribosome dissociates from the RNA with a certain probability has been observed. Ribosomes often also read occasionally through a weak termination signal (opal) for producing a part of the polyprotein in small amounts. (iv) RNA replication is error-prone. During amplification of the RNA genome, base exchanges or frame-shifting base deletions will introduce stop codons at random. Such progeny RNA does not make full polyprotein but may contribute to the pool of proteins made by the viral RNA population in the cell. Genetically defective RNA thus gets packed into mature virions.

For longer genomes, this strategy would be too wasteful. The viral RNA makes a polyprotein, but only for the early (NS) genes. The rest of the message has to await replication: the minus strand is the template for the production of full viral strands ready for packaging as well as for a mRNA containing the structural genes on the 3'-terminal half of the genome. Several quite successful strategies for making several mRNAs from the minus strand intermediate have evolved (see section on positive strand RNA viruses).

When the infection is started with a minus strand RNA, several mRNAs can be transcribed immediately without waiting for replication to occur. A large group of RNA viruses called minus strand RNA viruses, all of them with larger genomes, follow such a strategy. Since there is no enzyme to make mRNA from the minus strand in the host cell, the virion must carry it with the genomic RNA. As with positive strand viruses, several substrategies have evolved, the larger genomes often having segmented RNA genomes. While most positive RNA viruses like the picornaviruses multiply in the cytoplasm, RNA viruses with more complicated mechanisms have to import and export RNA and proteins into and out of the nucleus, complicating the experimental investigation of replication mechanisms.

Prokaryotic RNA Viruses: Leviviridae

Prokaryotic RNA viruses have been an instrumental tool for the investigation of the expression of RNA genomes and the infection cycle of an RNA virus. All the necessary steps in infection, protein synthesis, replication and phage assembly have been studied in vitro. RNA coliphages are abundant ($>10^4$ plaque-forming-units/ml) in sewage and are thus used as indicators for the hygiene of the water supply. Their virions are icosahedra with diameters of about 23 nm containing one molecule of single-stranded RNA with a nucleotide chain length of 3500 (group A) or 4200 (group B) coding for four genes that suffice for all necessary viral functions (Figure 2.4). The phage "head" is composed of 180 identical coat proteins (C), in some phages covalently linked by S-S bridges. Recognition of the host receptor, the F-pilus, is effected by one copy of the maturation protein, the product of the genes A or A2.

Upon entry into the host cytoplasm, translation proceeds from 5' × 3', while replication must start at the 3' terminus of the template and move to 5'. A clash is avoided by making sure that an RNA in the process of translating cannot replicate. Similarly, premature packaging of an RNA would be highly deleterious. Production of different amounts of gene products and a time regulation (phage clock) is effected by modulating the access of the "processors" for protein synthesis, replication and structural proteins to their specific binding sites on the RNA.

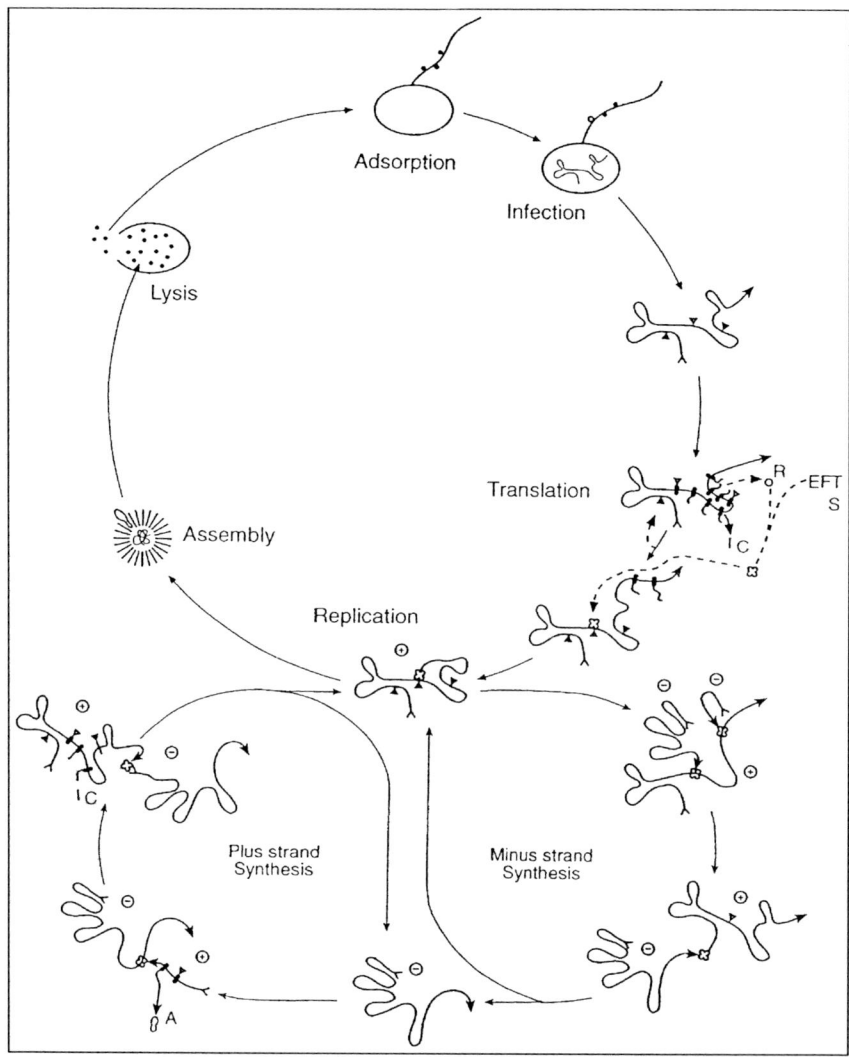

Fig. 2.5. Cartoon of the infection cycle of the bacteriophage Qβ. >-5' terminus of the RNA; → 3' terminus of the RNA; •ribosomes; ⊗replicase; ⌄ ribosome binding site avalaible; π ribosome binding site unavailable. At the beginning of the infection cycle, a levivirus particle binds to the host pilus and its RNA enters the host cytoplasm. Only the ribosome binding site (RBS) of the C gene is accessible and biosynthesis of the product of the C gene (pC) initiates. The polyribosome formed opens up the RNA structure making accessible also the RBS of the R gene. The pR combines with the host subunits EF Tu and EF Ts and S1 to form RNA replicase. Replicase and ribosomes compete for binding to their directly adjacent binding sites. If a replicase has bound, it prevents access of ribosomes to the RBS, but the replication can only start when the ribosomes have left the RNA and the 3' terminus can bind to the replicase (Biebricher and Eigen, 1987). Replication starts producing minus strands which bind neither ribosomes nor coat protein, but are excellent templates for the production of plus strands. Accumulated coat protein dimerizes and binds to a specific site (cb) preventing access of ribosomes to the RBS of the R-gene. Binding of coat protein dimer thus shuts off replicase biosynthesis and initiates capsid formation. Access to the RBS of the A2-gene is only possible on nascent replicating plus strand. Balancing of the protein synthesis, replication and capsid formation is

The genetic details of the infection cycle are shown in Fig. 2.5 (Weissmann, 1974; Biebricher and Eigen, 1988; Eigen et al, 1991).

Positive Strand RNA Viruses

Although retroviruses do contain an RNA genome of positive polarity, we refer to positive strand RNA viruses as those which replicate via a negative sense RNA but do not use cDNA as a replicative intermediate. Positive strand RNA viruses are abundant in animals and plants. They may use one or several messenger RNAs for synthesis of their proteins. The simplest representatives of this group are the animal picornaviruses (poliovirus, rhinoviruses, foot-and-mouth disease virus) with genomes of about 8000 residues and a single messenger RNA type of molecule in infected cells (Fig. 2.6). Next in complexity are the alphaviruses, a group of positive strand RNA viruses which produce a subgenomic RNA for the synthesis of structural proteins (Fig. 2.7). At the end of the RNA virus complexity scale, the coronaviruses have a genome of 27,000 to 32,000 nucleotides and synthesize a number of nested messenger RNAs in the infected cells.

In the positive strand RNA viruses some messenger RNAs are translated into polyproteins which are then processed by viral or cellular proteases to produce the mature, functional structural and nonstructural proteins. Several positive-strand RNA viruses use a Cap-independent internal initiation of translation through a structurally complex element termed the "internal ribosome entry site" (IRES) (Stewart and Semler, 1997; Lemon and Honda, 1997). Structural (S) proteins are those incorporated into viral particles, and nonstructural (NS) proteins are those involved in intracellular processes of viral multiplication. Picornavirus RNA is translated into a single polyprotein spanning all of the protein-coding potential of its genome. In contrast, coronaviruses use one messenger for each specific viral protein. These functional differences are relevant from an evolutionary point of view since they affect the tolerance of the virus to lethal or debilitating mutations. In picornaviruses, any mutation producing a termination codon within the open-reading frame of the polyprotein will be lethal. In a random nucleotide sequence one termination codon every 64 nucleotides is expected. Picornaviruses have a remarkably low number of infectious particles relative to physical particles (in the order of $1:10^3$ to $1:10^4$). It is believed that lethal mutations, together with assembly defects, and reinfection inefficiencies may contribute to such a low specific infectivity. Lethal mutations may be complemented by *trans*-acting functions supplied by a competent genome replicating in the same cell. The number of *trans*-acting versus *cis*-acting functions during replication of a viral genome is probably one of the factors influencing the possibility of sustaining the replication of defective viral genomes. Although complementation is a means to maintain highly debilitated genomes in a replicating ensemble, it is obvious that any cloning event resulting in separation of competent and defective genomes will result in extinction of the latter. This is one of many important consequences that genetic bottlenecks (replication initiated by one or few viral genomes) have in the evolution of viral quasispecies (Chapter 7).

thus done by competition of the ribosomes, replicase and coat protein for binding to RNA. Ribosomes are initially in large excess and RNA synthesis is not detectable in the first 14 min, then in sudden burst phage RNA and protein synthesis explosively increase until 20 min after infection (Eigen et al, 1991). Then replicase production ceases and production of viral RNA (nearly exclusively plus strand because replicase binds preferentially to the minus strand) and proteins proceed at maximum speed. Mature phages accumulate then linearly in the interior of the host cell and are liberated by lysis of the host cell, about 50 min after infection. Host cell lysis is triggered by viral proteins, in some leviviruses by a special L gene (Beremand & Blumenthal, 1979), in others by the A2 gene product (Karnik & Billeter, 1985) that is also responsible for binding of the phage particles to the F-pilus.

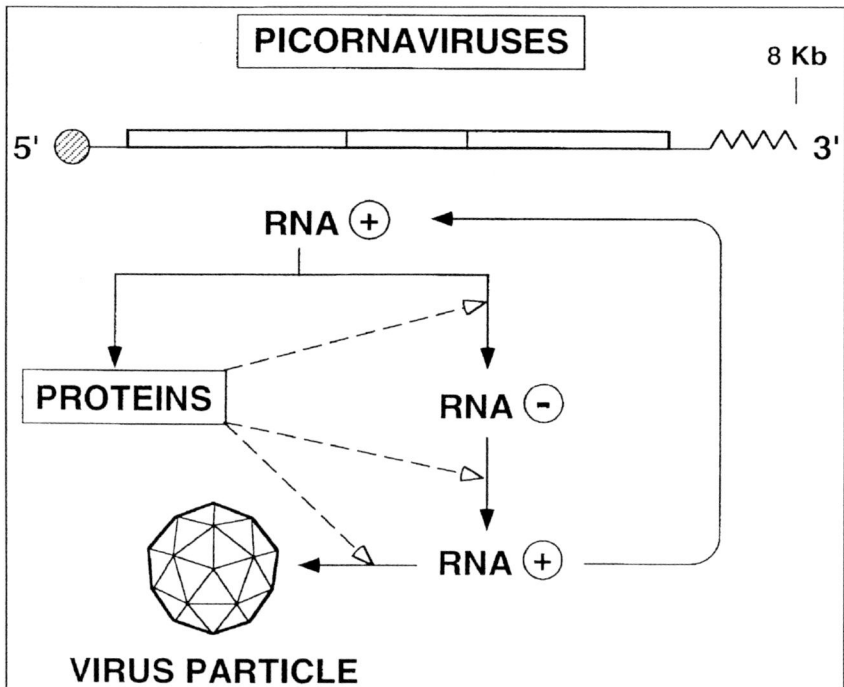

Fig. 2.6. Scheme of the picornavirus genome and its replication cycle. The genome is a RNA molecule of about 8 Kb with a protein covalently linked to the 5'-end of the RNA and a poly A tract at the 3' end. The genomic RNA acts as messenger RNA for the synthesis of viral proteins in the infected cell. The viral polymerase, together with host factors, catalyzes the synthesis of RNA of negative polarity which acts as a replicative intermediate in the form of partially double-stranded RNAs. Minus strand RNA serves as the template for the synthesis of RNA of positive polarity to serve as additional messenger RNA for the synthesis of viral proteins and also as progeny genomic RNA. The latter enters preformed, immature capsid shells to form the complete mature particles (virions) with icosahedral symmetry. The three-dimensional structure of a number of picornaviruses has bee elucidated by crystallographic methods.

Positive strand RNA viruses include many important animal and plant pathogens. Examples are poliovirus (the causative agent of poliomyelitis, a disease on its way to eradication) hepatitis C virus (HCV) (an infection which affects about 1% of the human population and that is associated with liver cirrhosis and hepatocarcinoma) and yellow fever virus, the agent of a reemerging tropical fever disease (Chapter 9).

Negative Strand RNA Viruses

The genomic RNA of this group of viruses has a polarity complementary to the messenger RNAs that encode the viral proteins. Negative strand RNA viruses include viruses with nonsegmented genomes that are termed mononegavirales. Examples are measles virus, rabies virus, and vesicular stomatitis virus. Other negative strand RNA viruses have a segmented genome such as the influenza viruses, and the increasingly important bunyaviruses and arenaviruses. The latter two groups are ambisense in that the same nucleotide sequence, read in its two polarities, encodes a distinct viral protein. By necessity, the replication cycle of negative strand RNA viruses must involve two different processes: transcription to produce functional messenger RNAs of positive polarity (and then their encoded proteins) and replication to yield

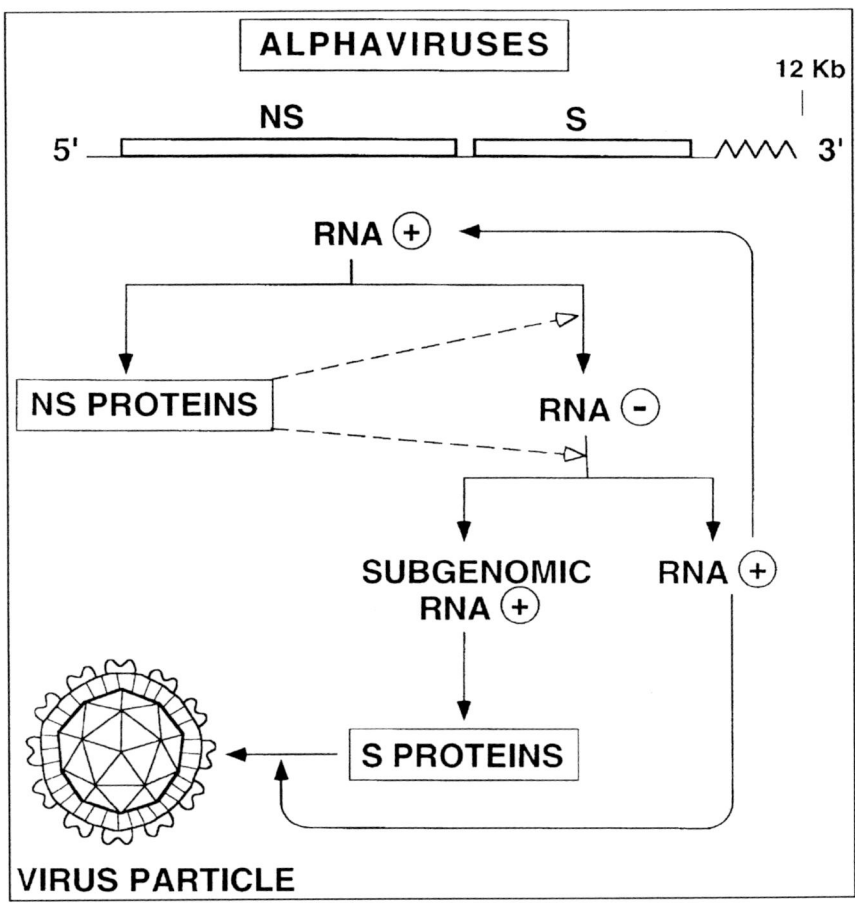

Fig. 2.7. Scheme of an alphavirus genome and its replication cycle. The polyadenylated positive strand RNA is about 12 Kb in length, and it encodes nonstructural (NS) and structural (S) proteins. The genomic RNA serves as messenger RNA for the synthesis of NS proteins. The latter catalyze the synthesis of negative sense RNA that serves as template for the synthesis of subgenomic messenger RNAs and full length progeny RNAs. Subgenomic messenger RNA directs the synthesis of S proteins for virus assembly and formation of the enveloped particles.

negative sense RNA for encapsidation into viral particles (Fig. 2.8). The nucleocapsid (located inside the viral envelope) includes the RNA-dependent RNA polymerase (the product of gene L). This enzyme, together with additional viral and cellular proteins, catalyzes the synthesis of multiple messenger RNAs as well as full-length positive strand RNA which serves as a replication intermediate. Each messenger RNA generally directs the synthesis of one viral protein rather than the synthesis of a polyprotein as in the case of positive strand RNA viruses (compare Figs. 2.6-2.8).

Influenza virus is the most typical segmented (or multipartite) RNA virus and one of the best studied from many points of view. Human influenza virus type A includes eight RNA segments which produce a total of ten proteins, two of them as a result of splicing events. Nine virus-coded proteins are found in virus particles. Important concepts in evolutionary virology and viral pathogenesis were first established with influenza viruses. Among them, the distinction

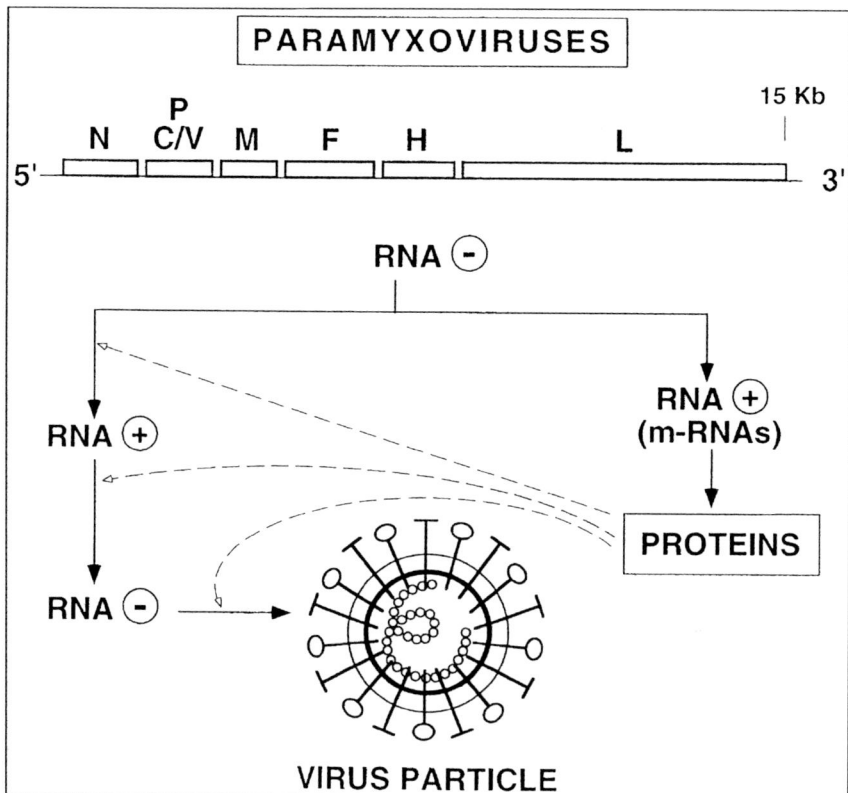

Fig. 2.8. Scheme of a paramyxovirus genome and its replication cycle. The genomic RNA is 15 Kb in length and of negative polarity. In consequence the genomic RNA must be transcribed by a virion transcriptase to produce messenger RNAs of positive polarity for the synthesis of viral proteins. The latter mediate genome replication via a full length intermediate of positive polarity and virus assembly to produce enveloped progeny particles.

between antigenic shift and antigenic drift as a result of two different molecular mechanisms of variation: genome segment reassortment and mutation, respectively (Chapters 3 and 8).

It has been suggested that segmentation of an RNA virus genome may have been selected as an adaptive strategy to minimize the effect of lethal mutations in progeny viruses. Whatever the reason, segmentation confers to this and other virus groups, a great genetic flexibility because of the possibility of modulating, optimizing, or excluding genome segment constellations, enriched in diversity by the error-prone replication of each individual segment.

Multipartite (Segmented Genome) RNA Viruses

In addition to the human influenza viruses, a considerable number of animal and plant RNA viruses have segmented genomes. Particles of the animal reoviruses include 10-12 segments of double stranded RNA. Several plant viruses with single stranded RNA of positive polarity have their genomes split into two or three pieces which are encapsidated into different particles. The tobraviruses and comoviruses, represented by tobacco rattle virus and cowpea mosaic virus, respectively, have two RNAs encapsidated in separate particles. Each RNA encodes different viral proteins. Bromoviruses contain three different genomic RNAs of 3.2, 2.8 and

2.1 Kb which encapsidate into separate particles with icosahedral symmetry. Replication of bromovirus has been extensively studied by P. Ahlquist and colleagues who have devised elegant methodology exploiting yeast genetics to identify host factors needed for viral replication (Sullivan and Ahlquist, 1997). With a related virus, cucumber mosaic virus, an in vitro replication system with a functional viral RNA-dependent RNA polymerase was developed (Hayes and Buck, 1990). The replicase activity was dependent on added template and it supported the synthesis of both positive and negative strand RNA progeny. Plant viruses are involved in considerable genetic variation, competition, selection and long-term stability of consensus sequences (van Vloten-Doting and Bol, 1988; Roossinck, 1997) which are a frequent hallmark of viral quasispecies. The vast majority of all known plant viruses are ordinary RNA viruses (riboviruses) which do not involve DNA intermediates in their replication.

Subviral Agents

Viroids are plant pathogens that consist of circular RNA of 200-400 nucleotides in length which do not express any protein and do not require a helper virus for replication (Diener, 1996; Flores et al, 1997). They show a rodlike structure as a result of extensive intramolecular base pairing. Viroid RNA is replicated by the RNA polymerase II found in the nucleus of plant cells by a rolling circle mechanism. Viroids are diverse in size, nucleotide sequence and phenotypic (disease) manifestations. One group forms an autocatalytic, RNA hammerhead structure with properties of a ribozyme. The self-cleaving activity generates monomeric, mature viroid RNA as a processing product of multimers produced during replication. It is believed that viroids may be relics of a primitive RNA world (Diener, 1989). Virusoids are quite similar to viroids, except that they depend on a helper virus for replication.

Satellites are RNA molecules about 400-2000 residues long which, contrary to viroids, are dependent on a helper virus for their replication (Kurath and Robaglia, 1995). They are generally associated with plant RNA viruses and they may either encode their own coat protein for encapsidation or use the coat of the helper. Satellite RNAs may modulate the symptoms caused by the helper virus, but their genomes show no detectable sequence similarity with their helper viruses (Vogt and Jackson, 1999).

An interesting replicon of animals which shares some properties with viroids and satellites is the 1700 residue-long hepatitis delta virus (HDV) or δ agent (Taylor, 1996, 1999; Robertson and Neel, 1999), a representative of the type of molecules that might have established possible links between the RNA and DNA worlds. Similarities of HDV with viroids include: the RNA is a covalently closed circular molecule, it collapses into a compact rodlike structure via extensive base pairing, it has a ribozyme activity, and its replication is mediated by the host RNA polymerase II. HDV replicates with the participation of several host transcription factors and a protein antigen (δAg-S) encoded by HDV. The ability to encode protein (two forms of the delta antigen, S and L) distinguishes this helper-dependent particle from viroids. Also, HDV is totally dependent on its helper virus, the hepadnavirus hepatitis B virus (HBV) for its replication. The envelope of HDV contains proteins encoded by the helper HBV, but the internal nucleocapsid includes several molecules of delta antigen. The genomic RNA is of negative polarity and during positive strand synthesis two forms of messenger RNA are synthesized as a result of an RNA editing event, a post-transcriptional modification utilized by some RNAs which expands their coding capacity (Benne, 1996). The HDV messenger RNAs direct the synthesis of the two forms of delta antigen which perform distinct and essential roles in the life cycle of this subviral satellite of HBV.

Defective Viral Genomes

Most viruses may give rise during their replication to defective viral genomes which are totally dependent on a helper virus for their replication (Holland, 1990; Vogt and Jackson,

1999). Some defective viruses may interfere with the replication of the standard, helper virus because of a competition for replication enzymes and protein factors. This establishes processes of coevolution between viruses and their defective counterparts. One of the most thoroughly studied systems is that of vesicular stomatitis virus (VSV) and its defective-interfering (DI) particles. There are several types of VSV DI RNAs, but they are generally highly deleted (subgenomic) forms of infectious VSV. They retain terminal genomic elements with the information necessary to be replication-competent. However, DI RNAs of VSV are replicated by *trans*-acting functions supplied by the standard VSV because the genomic segments encoding such functions are often absent or defective in the DI RNAs. The VSV-DI system was one of the first with which an extensive dynamics of mutation, competition and selection was revealed, and led Holland et al, (1982) to emphasize the biological implications of rapid evolution of RNA genomes. DI's can only be amplified efficiently when VSV is passaged at high multiplicity of infection (m.o.i., the number of infectious virus added per cell in an infection assay). This is expected because complementation of defective functions necessitates coreplication of DI and helper VSV in the same cell. When DIs are replication-competent, upon serial passages at high m.o.i. they can reach high proportions of total virus production because their replication occurs at the expense of helper virus replication. Dominance of DIs, however, must inevitably be limited by the concentration of complementing proteins that can be supplied by decreasing levels of helper virus. Before this point is reached, if VSV helper virus mutants with decreased sensitivity to DI interference arise, they will dominate over the previous DI's, until new mutant DI's are generated. This establishes a cyclic evolutionary alternation of VSV and its corresponding DIs that illustrates the continuum of change in rapidly evolving VSV-DIs (Holland et al, 1982; Holland 1984; 1990). This is an interesting example of regulatory feedback in which dominance of two alternative forms of replicons (a standard RNA and its DI) depends on the periodic generation of mutations in both as they affect the relative replicative fitness of the two entities.

Retroelements and Retroviruses

A very wide variety of cellular elements (collectively known as retroid agents) and viruses employ a reverse transcription step to copy RNA into DNA, a process which was not included in the so-called "dogma of molecular biology" which assumed a flow of genetic information from DNA to RNA to proteins (Chapter 1). For an historical account of the development of the DNA provirus hypothesis and its implications for carcinogenesis see Cooper et al, (1995). Retrotransposons share with retroviruses the copying of RNA into cDNA and the integration of cDNA into the host DNA (Coffin, 1996; Mc Clure, 1999). Most of them, however, do not form infectious particles, a hallmark of retroviruses. Because of its requirement for replication, all retroelements and retroviruses have a terminal repeat (TR) at both ends of the RNA. This TR is variable in length when different elements are compared. For example, the Ty element of *Saccharomyces cerevisiae* is 5,900 base pairs long with a 340 bp long TR. In contrast, the most complex retrovirus described to date, the human immunodeficiency virus (HIV) is about 9,700 residues long and it has a 650 residues long terminal repeat (LTR). The LTR of HIV is a complex element which includes, from its terminus, first a regulatory region (which contains recognition sequences for cellular transcription factors), a promoter region (to allow transcription by the cellular RNA polymerase II) and the regulatory element TAR which is essential for the activation mediated by a viral protein termed Tat (Chen et al, 1995; Levy, 1998).

Retroviruses include in their replication cycle (Fig. 2.9) two error-prone steps: DNA-dependent RNA synthesis with error rates of 10^{-5} to 10^{-6} misincorporations per nucleotide and round of copying (Rosenberger and Hilton, 1983; Blank et al, 1986), and reverse transcription also with about 10^{-5} to 10^{-6} misincorporations, deletions and frame-shifting events per nucleotide and round of copying (Temin, 1993). In addition, reverse transcriptase, in its

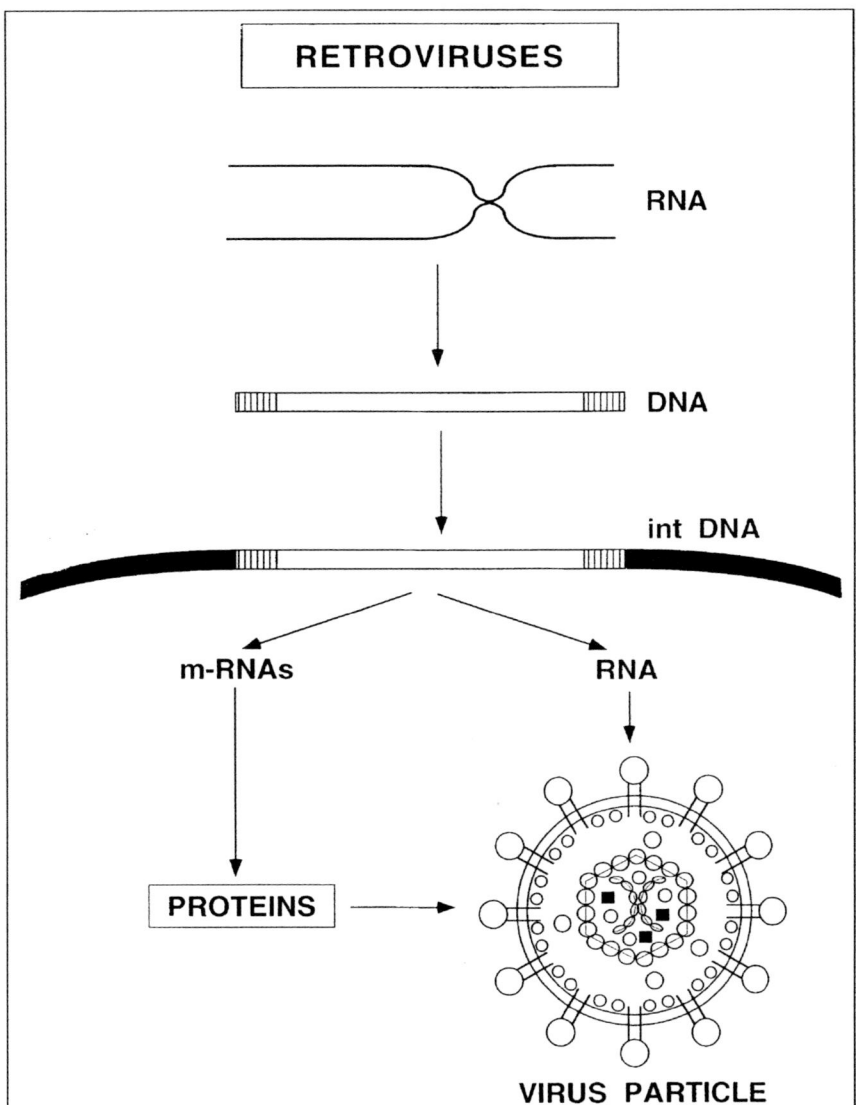

Fig. 2.9. Scheme of the retroviral life cycle. Retroviruses have the unique feature among viruses of including two copies of genomic RNA of positive polarity into each virion (diploid genome). Reverse transcriptase present in virus particles copies the RNA genome into a double stranded DNA which is integrated in cellular DNA. It is from this integrated DNA that viral prem-RNAs, m-RNAs and proteins are synthesized to produce enveloped progeny particles.

copying of the diploid retroviral genome (each virion includes two genomic RNA molecules with the same genetic contents), must jump from one RNA template molecule onto the other in order to complete a full length cDNA copy of the genome (Coffin, 1996; Coffin et al, 1997). This strand transfer is probably the molecular basis of an ability of retroviruses to produce recombinant genomes with high frequency. Recombination occurs preferentially between the two parental RNAs present in the same particle, as elegantly shown by H. Temin and associates (Temin, 1991). The structure and catalytic properties of reverse transcriptase and of other nucleic acid polymerases are discussed in Chapter 3.

The family *Retroviridae* includes a broad variety of human and animal pathogens, such as the lentiviruses HIV-1, HIV-2, and their simian counterparts (SIV), and the equine infectious anemia virus, among others. Many important observations on the quasispecies nature and population dynamics of RNA elements, involving both experimental and theoretical studies, have been made with retroviruses, in particular with HIV-1 in cell culture and in vivo (Coffin et al, 1997).

Hepadnaviruses

This is an unusual group of viruses with a gapped (partially double stranded) DNA genome and several unique replication features such as a reverse transcription step in its life cycle (Ganem, 1996; Hu and Seeger, 1997). Salient features of the hepadnavirus HBV are a small DNA genome of 3.2 Kb with several open reading frames (ORFs) including a DNA polymerase (Pol) with reverse transcriptase activity. Following entry into the cell, nucleocapsids are transported to the nucleus where synthesis of covalently closed circular DNA (cccDNA) takes place. This is then transcribed by the cellular RNA polymerase II, and the m-RNAs are translated to give the viral proteins (Fig. 2.10). cccDNA serves also as template for the synthesis of a full length RNA termed the pregenomic RNA which is encapsidated into core particles together with Pol. In these particles, synthesis of less-than-full length minus strand DNA takes place with concomitant RNA degradation by the RNase H activity associated with Pol. Also encoded as a part of Pol is a protein covalently linked to the 5'-end of the DNA which serves as a primer for plus strand DNA synthesis, a feature shown also by bacteriophage $\phi29$ or the adenoviruses. Assembly results in enveloped particles with the surface (s) antigen as an important immunological marker.

Expression of surface, core and precore antigens are important markers of diseases caused by HBV in humans. Even before disease symptoms appear, the s antigen can be detected, and antibodies against s are indicative of a good immune response and may be protective against reinfection. Precore and core antigens are indicative of active viral replication, with up to 10^{12} particles per ml of blood found in infected individuals at this stage. Precore antigen positivity is often associated with hepatic lesions. The X protein (Fig. 2.10) is a transcription *trans*-activator that activates cellular signal transduction pathways and binds to proteins involved in cellular DNA repair. It is believed that X may be one of the factors which lead to hepatocarcinoma, one of the dreaded consequences of HBV infections. HBV may cause acute disease and also chronic infections that enhance the probabilities of transmission of the virus from infected individuals to susceptible ones for prolonged time periods (Ewald, 1994). This brief description of HBV illustrates how, as in the case of many other viruses, viral gene expression and its detection are important for disease diagnosis and prognosis, and how expression of some viral genes may be the direct cause of some types of pathology.

The extraordinary life cycle of HBV exemplifies also how functional modules can assemble to fulfill the aim of replication with remarkable exploitation of highly compact genetic information. HBV uses overlapping open-reading frames (ORFs), specific posttranslational modifications of functional significance, and multifunctionality of Pol and of other specific genomic sites. For example, a stem loop termed ε, located within terminal redundant repeats found in pregenomic RNA, is involved in RNA encapsidation and reverse transcription (Ganem, 1996).

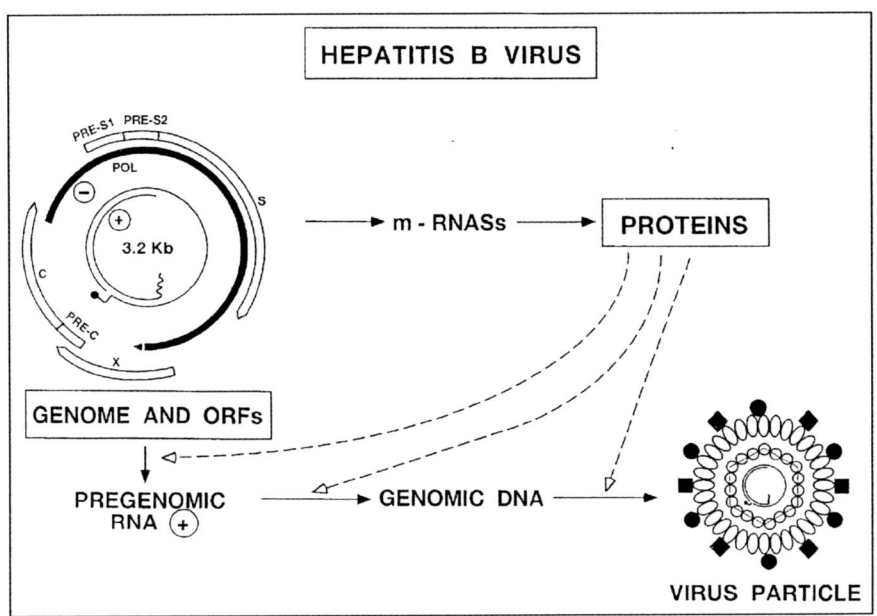

Fig. 2.10. Scheme of the hepatitis B virus genome, open-reading frames, and replication cycle. The 3.2 Kb circular genome is partially double stranded with a protein covalently linked to the 5' end of the DNA (black dot). Overlapping reading frames are indicated as open and filled partial circles. Proteins encoded are several forms of surface antigen(s), core (c), polymerase (Pol) and protein X. Covalently closed DNA is the template for the synthesis of mRNAs which are translated into viral proteins. A full length pregenomic RNA is encapsidated into core particles and retrotranscribed by Pol to yield the genomic DNA. Cores bud into membranes of the endoplasmic reticulum, and the enveloped virus particles are secreted from the cell.

Disparate Life Cycles, Common Survival Strategies

The variety of multiplication strategies outlined in preceding paragraphs must not hide the fact that the main objective, the very essential feature for which most viruses have been selected, is replication for long-term survival. It is reasonable to postulate that viruses have inherited from the simple replicons of a primitive RNA world their main purpose: to produce and maintain more of themselves. Abundance and stability are features that must have been strongly selected for early replicons. Most present day viruses are best known for their potential to cause disease. Yet their primitive precursors were most probably selected for their ability to spread among cells. In so doing they must have contributed blindly to the adaptation of cells to changing environments, via perturbation and restructuring of cellular genomes. Positive selection for replication of viral genomes resulted in potential beneficial but also detrimental effects for an increasingly organized cellular world. Retroid agents are a clear example (Britten, 1997; McClure, 1999): Mobile genetic elements can provide regulatory sequences for host gene expression, may be involved in repair of breaks in DNA, in telomere maintenance, and may have contributed to speciation events. In sharp contrast, some retroid agents have been associated with human genetic disease through insertional mutagenesis of human genes. Examples are some types of hemophilia and Duchenne muscular dystrophy (reviewed by McClure, 1999).

Potential beneficial or detrimental effects of an RNA genetic element can be observed with some plant satellite RNAs. Some satellite RNAs reduce the replication levels of helper virus and attenuate disease symptoms in infected plants. Other satellite RNAs have an opposite, disease-enhancing effect (Roossinck et al, 1992). In some cases, mutation can convert a benign satellite into a pathogenic one (Paulukatis and Roossinck, 1996). The disparate phenotypic effects of RNA genetic elements are necessarily associated with their capability to replicate or to interfere with replication of other genomes.

Evolution of a virus within an individual organism may often be short-sighted in that the virus evolves in response to the particular constraints which individual hosts pose to its multiplication (Levin and Bull, 1994). However, success in one host without maintenance of a good probability for transmission to new hosts would entail an evolutionary dead-end for the virus (Ewald, 1994). The complex array of interactions that viruses unavoidably establish with multitudes of organisms (and organs, tissues and cells within organisms) tends to obscure the ultimate trait for which they have been (and continue to be) selected: survival. Disease is only a by-product of their "raison d'être". Thus it should not be surprising that some traits related to the need for survival are shared by viruses of all groups discussed in preceding paragraphs, no matter how disparate their detailed replication pathways. Error-prone replication, with reservoirs of genetic and phenotypic variants, and frequently also a rapid turnover of virions and of infected cells, can be important features facilitating survival and persistence (Domingo et al, 1998).

The adaptability of viruses within infected organisms is directly linked to difficulties for medical control of the diseases they cause. HIV-1, HCV and HBV produce viral particles in infected humans with a half-life of a few hours. It is estimated that between 10^{10} and 10^{12} new particles are produced in each infected individual each day. In all three cases, mutant spectra contain phenotypic variants that allow the viruses to escape externally applied or intrahost selective pressures, be it inhibitor drugs or antibodies or cytotoxic T-lymphocytes (Domingo, 1989; Domingo and Holland, 1992; Novella et al, 1995; McMichael and Phillips, 1997). Most clinicians have agreed that combination drug therapy (to achieve maximum reduction of viral replication and to avoid the selection of multiresistant mutants) is probably the best approach for viral therapy. Yet, the pernicious adaptability of viruses allows them to escape multilateral pressures with varying degree of success. The result is that therapeutic strategies are periodically reexamined and alternatives sought. Because of the fundamental links between the quasispecies structure of RNA viruses and the difficulties for viral disease control, this topic is considered in greater detail in Chapter 8.

Selection for abundance and for resistance to externally applied selective pressures is by no means unique to viral pathogens. It is the essence of the evolution of life on earth. From the point of view of human disease it underlies difficulties to control bacterial, fungal and other parasitic diseases. Cancer must be viewed as a mutation-driven process in which invasiveness and metastasis are the result of adaptation of heterogeneous cell populations to different microenvironments (Nicolson, 1987; Temin, 1988; Ionov et al, 1993) (see also Chapter 7). In all these cases, the underlying Darwinian evolutionary principles are the same.

Early Evidence of RNA Virus Variability

Before nucleotide sequencing techniques were available, virologists already had indications that RNA viruses were genetically unstable. Common observations were that wild type revertants were frequently generated upon passage of mutant viruses, and that a considerable proportion of mutants were found in wild type preparations of viruses that had not been subjected to any chemical mutagenesis treatment. In 1940 L.O. Kunkel showed that as much as one in 200 lesions on tobacco leaves were induced by some variant form of tobacco mosaic virus. During the 1940s, F.M. Burnet studied the changes in agglutinating activity of influenza viruses and stated that attenuation of a virus upon passage in an unusual host "must necessarily be something of an exercise in population genetics". High frequencies of spontaneous mutants or wild type revertant viruses were found by A. Granoff working with Newcastle disease virus, B.N. Fields and W.K. Joklik with reovirus, and R.C. Valentine and colleagues with bacteriophage Qβ, among other studies. Pelham and colleagues, referring to an earlier paper of their group, wrote in 1970: "The use of tomato varieties protected only by single genes for resistance to TMV (Tobacco mosaic virus), without taking adequate precautions to minimize the exposure of the host to the pathogen, is the surest way of selecting and establishing new strains of the pathogen and so wasting the few available genes for resistance" (Pelham et al, 1970). This reflects a general concept that was to be dramatically verified two decades later with the inadequate use of antiviral agents and antibiotics. Early results on phenotypic variation of RNA viruses have been reviewed by Domingo and Holland (1988).

Many observations with animal and plant RNA viruses suggested considerable genetic instability when attempting to apply to these pathogens the approaches of mutant preparation and characterization developed for bacteria and later for yeast. The true magnitude of the problem became apparent only when nucleotide sampling techniques started being applied to RNA phages. After the first bacteriophage RNA sequences were obtained, Charles Weissmann and his colleagues wrote in 1973: "An apparently phenotypically homogeneous phage stock might contain multiple variants at various sites on the RNA". However, the concept of deep genetic heterogeneity was not generally accepted, perhaps partly due to the influence of the notion of genetic polymorphism in population genetics which excluded as polymorphic alleles those variant loci found only once (see Chapter 9). Even in the early 1980s (several years after a number of papers on viral quasispecies had already been published) a common view was that the cloning and expression of the relevant subtype hemagglutinins of human influenza virus (or surface antigens of any pathogen for that matter) would provide the requisite battery of antigens needed for vaccine production, depending on the circulating strain. The concept of antigenic variation was well established but the notion of mutation, competition and selection as a continuous source of variation even in virus within a single infected cell was not even contemplated. Mutant generation is mainly a property of the replication enzymes as they recognize and copy viral nucleic acids. Other sources of RNA mutation exist: they include such chemical changes as oxidation, deamination, depurination, depyrimidination, etc., as well as several forms of radiation damage.

References

1. Alkhatib G, Combadiere C, Broder CC et al. CC CKR5: A RANTES, MIP-1 alpha, MIP-1 beta receptor as a fusion cofactor for macrophage-tropic HIV-1. Science 1996; 272:1955-58.
2. Benne R. RNA editing: How a message is changed. Curr Opin Genet Dev 1996; 6:221-31.
3. Biebricher CK, Eigen M. Kinetics of RNA replication by Qβ replicase. In: Domingo E, Holland JJ, Ahlquist P, eds. RNA Genetics, vol. 1. Boca Raton Fl: CRC Press Inc. 1988; 1-21.
4. Blank A, Gallant JA, Burgess RRet al. A RNA polymerase mutant with reduced accuracy of chain elongation. Biochemistry 1986; 25:5920-5928.
5. Britten RJ. Mobile elements inserted in the distant past have taken on important functions. Gene 1997; 205:177-82.
6. Cao W, Henry MD, Borrow P et al. Identification of α-dystroglycan as a receptor for lymphocytic choriomeningitis virus and Lassa fever virus. Science 1998; 282:2079-2081.
7. Chen ISY, Koprowski H, Srinivasan A, Vogt PK eds. Transacting functions of human retroviruses. Curr Top Microbiol Immunol vol. 193. Berlin: Springer-Verlag, 1995.
8. Coffin JM. *Retroviridae*: The viruses and their replication. In: Fields BN, Knipe DM, Howley PM et al, ed. Fields Virology. Philadelphia: Lippincott-Raven 1996:1767-1847.
9. Coffin JM, Hughes SH, Varmus HE. Retroviruses. New York: Cold Spring Harbor Laboratory Press, 1997.
10. Cooper GM, Temin RG, Sugden B. The DNA provirus. Howard Temin's scientific legacy. Washington DC: American Society for Microbiology; 1995.
11. Crawford EM, Gesteland RF. The adsorption of bacteriophage R17. Virology 1964; 22:165-167.
12. Diener TO. Circular RNAs: Relics of precellular evolution? Proc Natl Acad Sci USA 1989; 86:9370-4.

13. Diener TO. Origin and evolution of viroids and viroid-like satellite RNAs. Virus Genes 1996; 11: 47-59.
14. Diez J, Mateu MG, Domingo E. Selection of antigenic variants of foot-and-mouth disease virus in the absence of antibodies, as revealed by an in situ assay. J Gen Virol 1988; 70:3281-88.
15. Domingo E. RNA virus evolution and the control of viral disease. Prog Drug Res 1989; 33:93-133.
16. Domingo E, Baranowski E, Ruiz-Jarabo CM et al. Quasispecies structure and persistence of RNA viruses. Emerging Infectious Diseases 1998; 4:521-527.
17. Domingo E, Holland JJ. High error rates, population equilibrium, and evolution of RNA replication systems. In: Domingo E, Holland JJ, Ahlquist P, eds. RNA Genetics, vol. III. Boca Raton: CRC Press Inc. 1988:3-36.
18. Domingo E, Holland JJ. Complications of RNA heterogeneity for the engineering of virus vaccines and antiviral agents. Genet Eng 1992; 14:13-31.
19. Eigen M, Biebricher CK, Gebinoga M et al. The hypercycle. Coupling of RNA and protein biosynthesis in the infection cycle of an RNA bacteriophage. Biochemistry 1991; 30:11005-11018.
20. Evans DJ, Almond JW. Cell receptors for picornaviruses as determinants of cell tropism and pathogenesis. Trends in Microbiol 1998; 6:198-202.
21. Ewald PW. Evolution of infectious disease. Oxford: Oxford University Press 1994.
22. Fields BN, Knipe DM, Howley PM et al, eds. Fields Virology. Philadeophia: Lippincott-Raven 1996.
23. Flint SJ, Enquist LW, Krug RM et al. Principles of virology. Molecular Biology, Pathogenesis and Control. Washington DC: American Society for Microbiology 2000.
24. Flores R, Di Serio F, Hernández C. Viroids: The non coding genomes. Seminars in Virol 1997; 8: 65-73.
25. Ganem D. *Hepadnaviridae* and their replication. In: Fields BN, Knipe DM, Howley PM et al, ed. Fields Virology. Philadelphia: Lippincott-Raven 1996:2703-37.
26. Hayes RJ, Buck KW. Complete replication of a eukaryotic virus RNA in vitro by a purified RNA-dependent RNA polymerase. Cell 1990; 63:363-8.
27. Holland JJ. Receptor affinities as major determinants of enterovirus tissue tropism in humans. Virology 1961; 15:312-26.
28. Holland JJ. Continuum of change in RNA virus genomes. In: Notkins AL, Oldstone MBA, ed. Concepts in viral pathogenesis. New York: Springer-Verlag 1984: 137-43.
29. Holland JJ. Defective viral genomes. In: Fields BN, Knipe DM, Chanock RM et al, ed. Virology. New York: Raven Press 1990:151-65.
30. Holland JJ, Spindler K, Horodyski F et al. Rapid evolution of RNA genomes. Science 1982; 215: 1577-85.
31. Hu J, Seeger C. RNA signals that control DNA replication in hepadnaviruses. Seminars in Virology 1997; 8:205-11.
32. Ionov Y, Peinado MA, Malkhosyan S et al. Ubiquitous somatic mutations in simple repeated sequences reveal a new mechanism for colonic carcinogenesis. Nature 1993; 36:558-61.
33. Kamen RI. Structure and function of the $Q\beta$ RNA replicase. In: Zinder ND, ed. RNA phages. Cold Spring Harbor Laboratory 1975:203-34.
34. Kunkel LO. Publ Am Assoc Adv Sci 1940; 12:22.
35. Kurath G, Robaglia C. Genetic variation and evolution of satellite viruses and satellite RNAs. In: Gibbs A, Calisher CH, García-Arenal F, ed. Molecular basis of viral evolution. Cambridge: Cambridge University Press 1995: 385-403.
36. Lai MMC. Cellular factors in the transcription and replication of viral RNA genomes: A parallel to DNA-dependent RNA transcription. Virology 1998; 244:1-12.
37. Lemon SM, Honda M. Internal ribosome entry sites within the RNA genomes of hepatitis C virus and other flaviviruses. Seminars in Virology 1997; 8:274-88.
38. Levin BR, Bull JJ. Short-sighted evolution and the virulence of pathogenic microorganisms. Trends in Microbiol. 1994; 2:76-81.
39. Levy JA. HIV and the pathogenesis of AIDS. Washington DC: ASM Press, 1998.
40. Longberg-Holm K, Philipson L. Early interactions between viruses and cells. New York: S Karger; 1974.

41. McClure MA. The retroid agents: Disease, function and evolution. In: Domingo E, Webster, RG, Holland JJ, eds. Origin and evolution of viruses. San Diego: Academic Press 1999: 163-95.
42. McLaren LC, Holland JJ, Syverton JT. The mammalian cell-virus relationship. I. Attachment of poliovirus to cultivated cells of primate and nonprimate origin. J Exp Med 1959; 109: 475-85.
43. McMichael AJ, Phillips RE. Escape of human immunodeficiency virus from immune control. Annu Rev Immunol 1997; 15:271-96.
44. McNight A, Dittmar MT, Moniz-Pereira J et al. A broad range of chemokine receptors are used by primary isolates of human immunodeficiency virus type 2 as coreceptor with CD4. J Virol 1998; 72:4065-71.
45. Mondor I, Ugolini S, Sattentau QJ. Human immunodeficiency virus type 1 attachment to HeLa CD4 cells independent and gp120 dependent and requires cell surface heparans. J Virol 1998; 72: 3623-34.
46. Nicolson GL. Tumor cell instability, diversification, and progression to the metastatic phenotype: From oncogene to oncofetal expression. Cancer Res 1987; 47:1473-87.
47. Novella IS, Domingo E, Holland JJ. Rapid viral quasispecies evolution: Implications for vaccine and drug strategies. Mol Med Today 1995; 1:248-53.
48. Palukaitis P, Roossinck MJ. Spontaneous change of a benign satellite RNA of cucumber mosaic virus to a pathogenic variant. Nature Biotechnology 1996; 14:1264-8.
49. Pelham J, Fletcher JT, Hawkins JH. The establishment of a new strain of tobacco mosaic virus resulting from the use of resistant varieties of tomato. Ann Appl Biol 1970; 65:293-97.
50. Robertson HD, Neel OD. Virus origins: Conjoined RNA genomes as precursors to DNA genomes. In: Domingo E, Webster RG, Holland JJ, eds. Origin and Evolution of Viruses. London: Academic Press 1999: 25-35.
51. Roossinck MJ. Mechanisms of plant virus evolution. Annu Rev Phytopathol 1997; 35:191-209.
52. Roossinck MJ, Sleat D, Paulukaitis P. Satellite RNAs of plant viruses: structures and biological effects. Microb Rev 1992; 56:265-79.
53. Rosenberger RF, Hilton J. The frequency of transcriptional and translational errors at nonsense codons in the *lac Z* gene of *Escherichia coli*. Mol Gen Genet 1983; 191:207-212.
54. Sattentau QJ. HIV gp120: Double lock strategy foils host defences. Structure 1998; 6:945-49.
55. Schneider-Schanlies J. Cellular receptors for viruses: Links to tropism and pathogenesis. J Gen Virol 2000; 81:1413-29.
56. Stewart SR, Semler, BL. RNA determinants of picornavirus Cap-independent translation initiation. Seminars in Virology 1997; 8:242-55.
57. Sullivan ML, Ahlquist P. *cis*-Acting signals in bromovirus RNA replication and gene expression: Networking with viral proteins and host factors. Seminars in Virology 1997; 8:221-30.
58. Taylor JM. Hepatitis delta virus and its replication. In: Fields BN, Knipe DM, Howley PM et al, eds. Fields Virology. Philadelphia: Lippincott-Raven 1996:2809-18.
59. Taylor JM. Human hepatitis delta virus: An agent with similarities to certean satellite RNAs of plants. Curr Top Microbiol Immunol 1999; 239:107-122.
60. Temin HM. Evolution of cancer genes as a mutation-driven process. Cancer Res 1988; 48:1697-701.
61. Temin HM. Sex and recombination in retroviruses. Trends in Genetics 1991; 7:71-4.
62. Temin HM. The high rate of retrovirus variation results in rapid evolution. In: Morse SS, ed. Emerging viruses. New York: Oxford University Press 1993:219-25.
63. van Vloten-Doting L, Bol JF. Variability, mutant selection, and mutant stability in plant RNA viruses. In: Domingo E, Holland JJ, Ahlquist P, eds. RNA Genetics, vol. 3. Boca Raton: CRC Press, Inc. 1988:37-51.
64. Vogt PK, Jackson AO, eds. Satellites and Defective Viral RNAs. Current Top Microbiol Immunol, vol. 239; 1999.
65. Weissmann C. The making of a phage. FEBS letters (Suppl.) 1974; 40:S10-S18.
66. Weissmann C, Billeter MA, Oodman HM et al. Structure and function of phage RNA. Annu Rev Biochem 1973; 42:303-28.
67. Wimmer E. Cellular receptors for animal viruses. Cold Spring Harbor Laboratory Press, 1994.

CHAPTER 3

Molecular Recognition and Replication Enzymes

The special chemistry of life is centrally governed by special biomacromolecules. These are built up as linear polymers from monomeric subunits: The key biopolymers are nucleic acids (deoxyribonucleic acid or DNA and ribonucleic acid or RNA) and proteins. Quite in contrast to industrial polymers there is a strong polarity in the strands: each monomer has two ends that can be distinguished by having different chemical groups: the 5' and 3' ends in nucleic acids and the amino (N-) and carboxy (C-) terminal ends in proteins.

Structure of Nucleic Acids and Restrictions to Variation

Nucleic acids are connected by phosphodiester bonds between the 5' hydroxyl groups of one nucleotide and the 3' hydroxyl of the next nucleotide. Nucleic acids, in particular DNA, can have very long chain lengths of several million: the *Escherichia coli* chromosome has 4×10^6 base pairs (bp) and the multiple chromosomes in a mammalian cell amount to about 3×10^9 bp. In addition to the four standard nucleotides, nucleic acids may contain also other odd or unusual nucleotides [1-methyladenine, 2-methylguanosine, 7-methylguanosine, 5-methylcytidine, pseudouridine, inosine, among others in transfer RNA (t-RNA), hydroxymethylcytosine in the DNA of T-even bacteriophages, and 5-methylcytosine in mammalian DNA]. These are usually not incorporated during normal nucleic acid biosynthesis, but rather they are formed by specific (enzymatic) chemical modifications of certain nucleotides in the chain.

Reproduction is an inherent property of nucleic acids when provided with monomeric substrates and a favorable biological and physical environment. In nature, only the genome is copied directly; all other parts of the cell must be synthesized by expression of the genetic information, by interconnected metabolic pathways, and by complex protein-protein, protein-lipid, protein-sugar and protein-nucleic acid interactions. Both nucleic acids and proteins may undergo a rather specific folding as a result of strong ionic interactions, hydrophobic interactions, weaker van der Waals contacts, and also long-range interactions. Chain-chain and solvent (water)-chain interactions play critical roles in the folding of biological macromolecules, as required to perform structural and functional roles. Interactions that involve a single biopolymer chain and that are formed early in the folding process are referred to as the secondary structure of the polymer. Further interactions between different chains or distant domains of the same chains conform the tertiary and higher order structures of the biopolymers. Such complex structures have been more thoroughly studied and characterized for proteins than for nucleic acids.

A number of concerted processes give rise to essential macromolecular structures such as ribosomes, mitochondria and membranes. The discovery of the double helical structure of the DNA with its regular base-pairing immediately suggested a mechanism for genome duplication (Watson and Crick, 1953). Hydrogen bonds can be established between a considerable

---

Quasispecies and RNA Virus Evolution: Principles and Consequences, by Esteban Domingo, Christof K. Biebricher, Manfred Eigen and John J. Holland. ©2001 Eurekah.com.

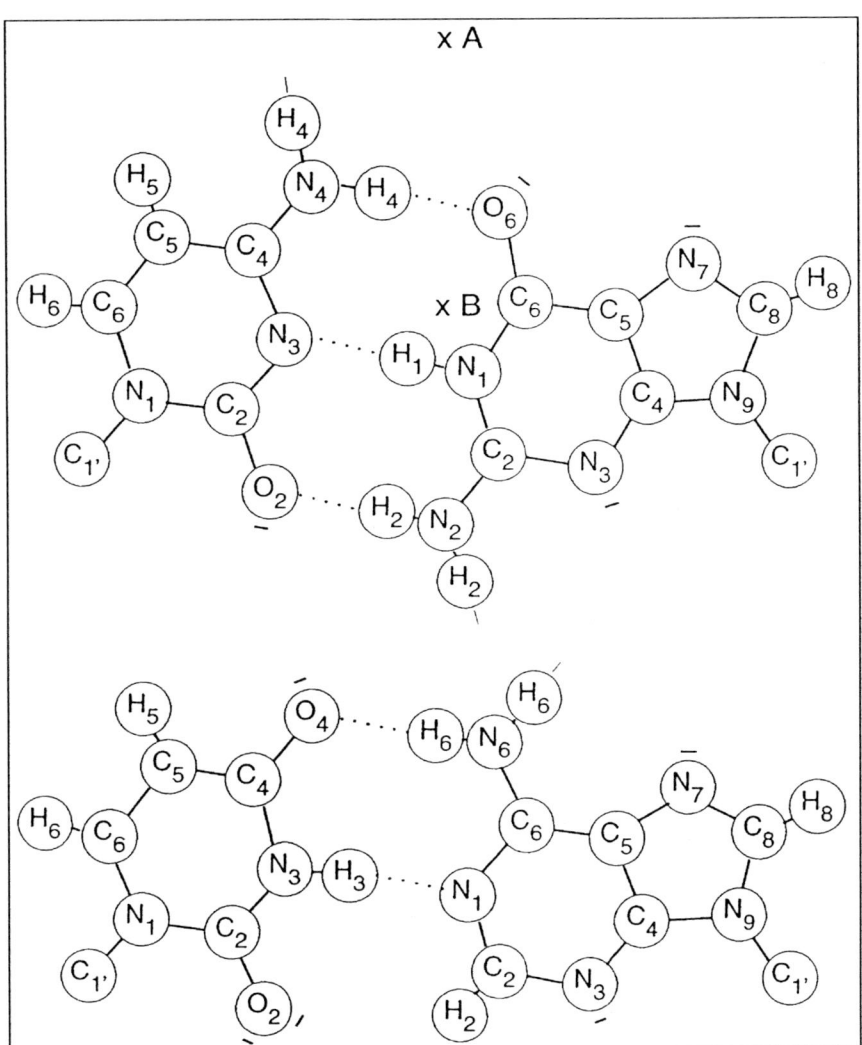

Fig. 3.1. Standard base pairs in nucleic acids. Note the three hydrogen bonds in G: C and the steric equivalence of the two base pairs. Residual hydrogen bond acceptor and donors of the base pairs in the major groove (upwards) and the minor groove (downwards) and the position of the helical axis of the B form (predominantly DNA) and the A form (RNA and DNA: RNA hybrids) are also shown.

number of identical or different bases, not only between the classical A-U (or A-T) and G-C (Fig. 3.1.). For example, the A:T base pairing by hydrogen bonding is not strongly preferred over the A:A or A:G pairings, the only exceptionally strong base pair being G:C. It is actually the steric equivalence of the standard Watson-Crick base-pairs and their pseudosymmetry that allows building of a regular structure without sequence restriction, and with the bases stacked one on another. This makes possible a stable, regular structure that leads to cocrystallization as large fibers of DNA double strands, even though the DNA may originate from different sources, and include different base sequences.

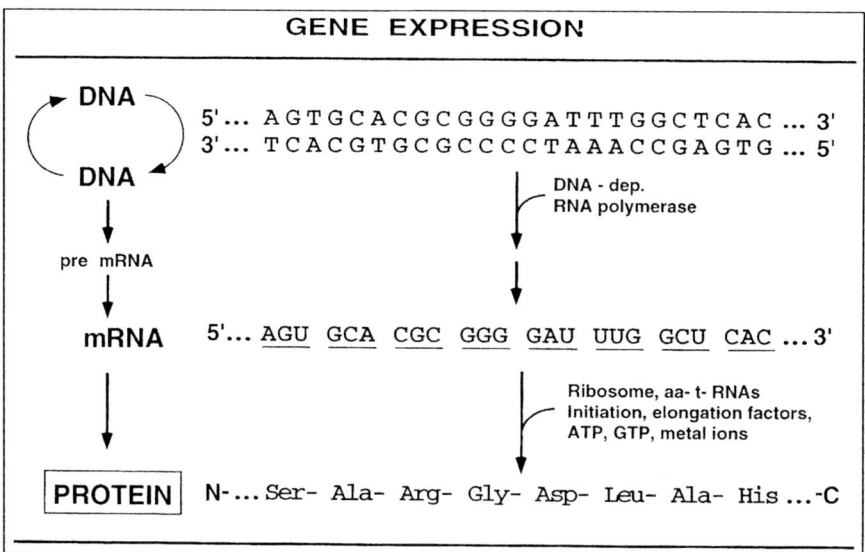

Fig. 3.2. Main steps in decoding the genetic material and gene expression. The four base symbol in double stranded DNA of opposite polarity is transcribed into messenger RNA (mRNA), often via splicing of precursor mRNAs. Triplets in the mRNA are translated into a polypeptide chain following the rules of the genetic code (see Fig. 3.3).

Nucleic acids are endowed with chemical properties suitable for the emergence of information and self-replication, a likely seed of living organisms. The four bases that constitute nucleic acids can encode a variety of symbols taking advantage of combinatorial properties. Linearly adjacent triplets in messenger RNA, a molecule directly copied from the double stranded DNA, direct the synthesis of a chain of amino acids according to the rules of the genetic code. A complex translation machinery with participation of RNAs and proteins achieves this decoding (Figs. 3.2 and 3.3). The stability of the double helical DNA ensures continuity of genetic information but not in an absolute manner. The interaction between complementary bases (Fig. 3.1) must necessarily have a limited strength and miscopying errors occasionally occur. This leads to replicative mutations, one of the sources of genetic variation and a driving force of evolution. In contrast to the genetic material, all other types of expressed RNA and proteins have a limited half-life. Errors in their synthesis do not generally cause permanent harm to progeny cells because nongenomic RNA and proteins are produced in a transient and renewable fashion. Recent evidence suggests that eukaryotic transcription may be subjected to some error-correcting mechanisms. RNA polymerases are capable of selectively removing misincorporated nucleotides during transcription in prokaryotic and eukaryotic cells (Thomas et al, 1998). The activity appears to be a 3' × 5' exonuclease which is activated by a number of cellular factors. The basis for the editing reaction has been attributed to a slow extension of a mispaired 3'-nucleotide at the elongation site. It is not yet clear whether such transcription-correcting activities may be operating in vivo. The fleeting existence of expression products provides the required flexibility of metabolism. Metabolites are needed discontinuously with regard to location and time.

Thus, the different degrees of stability of the macromolecules that constitute the genetic material and their expression products, as well as the combinatorial properties of the four bases that build DNA, are at the basis of the processes that we know as life. The same overall spatial configuration may embody myriads of distinct pieces of genetic information. It is DNA genomes

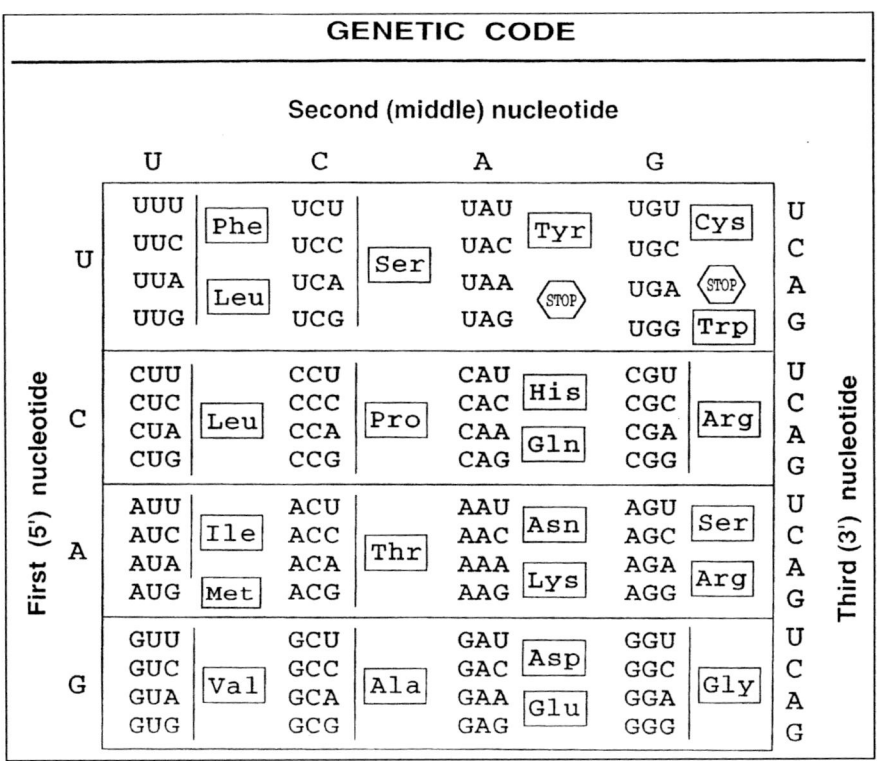

Fig. 3.3. Genetic code. Triplets may direct either the synthesis of an amino acid (boxed) or a stop signal that dictates protein termination. Note that the code is degenerate (several different triplets may encode the same amino acid) and that not all amino acid replacements require the same number and type of nucleotide substitutions.

composed of the same types of monomers (although in different number and order) that embody the information to generate an elephant, an herpes simplex virus, or a redwood tree.

With a few exceptions, DNA in nature is found in the double-stranded B form (Saenger, 1984). The 2'-hydroxyl of the RNA interferes with this conformation and, in consequence, double helices of RNA (and RNA:DNA hybrids) assume the A form (Saenger, 1984). RNA generally assumes the single-stranded form, but helical RNA regions as well as higher order RNA structures occur and often play important biological roles. The thermodynamic stability due to base-pairing in double strands can be calculated fairly accurately by adding the incremental contributions of base-pairs and their nearest neighbours to the ΔH and ΔS values (Wyatt and Tinoco, 1993; Nowakowski and Tinoco, 1997). The calculated stability can be experimentally verified by measuring the temperature at which the double helix becomes unstable and is melted into single strands. The stability of stem-loop RNA structures can be calculated by considering additional influences:

1. Noncanonical base-pairs contribute to the stability of the RNA helix, most notably G:U and G:A pairs.
2. Dangling (or wobbling) nucleotides, i.e., unpaired nucleotides next to a double-helical stem, contribute to stability via their stacking energy.
3. Stems can stack one on the other.

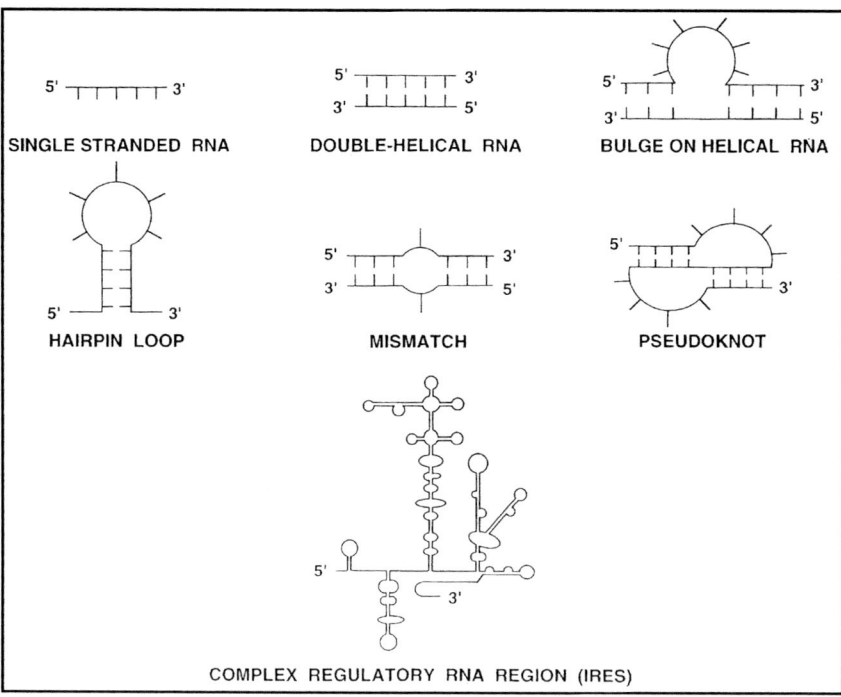

Fig. 3.4. Some types of recognized secondary and tertiary structures in RNA. The upper part shows six typical structural motifs on RNA. Pseudoknots are not planar structures as drawn. Rather they constitute three-dimensional tertiary RNA structures that play important biological functions. The lower part of the figure shows a constellation of secondary structure domains found in the internal ribosome entry sites (IRES) of some RNA viruses and a few cellular mRNAs. Again, this structure of about 450 residues is probably a complex three-dimensional structure with multiple tertiary interactions needed for specific interactions with proteins and for the translation-initiation activity of this element.

4. Interruptions of the double helix by looped-out single bases, bulges or loops decrease the stability.
5. There are tertiary structure contributions by pseudoknots, tetraloop sequences, loop-loop interactions, triple helices, loop-duplex interactions. These energetic contributions may lead to stable tertiary structures (Wyatt and Tinoco, 1993; Nowakowski and Tinoco, 1997; Deiman and Pleij, 1997) (Fig. 3.4).

For many types of RNA such as messenger RNA, ribosomal RNA, transfer RNA, ribozymes, and viral RNA, strong secondary and tertiary structures (not generally formed by random nucleotide sequences) have evolved and are often essential for these RNAs to perform some of their biological roles. Examples mentioned in the previous Chapter are the internal ribosome entry site (IRES) element needed for internal initiation of protein synthesis of several RNA viruses or the multifunctional ε element in pregenomic RNA of HBV. It has been shown for phage Qβ that evolution rapidly restores the strong secondary structure of its genomic RNA when it is disrupted by deletions or insertions introduced into the genome (Arora et al, 1996; Olsthoorn and van Duin, 1996). Poliovirus and other related picornaviruses contain structures at the 5'-and 3'-terminal genomic regions whose function in viral RNA replication is currently under intense investigation (Xiang et al, 1997). The 5'-untranslated region constitutes a cloverleaf-like structure required for the synthesis of positive sense RNA. Two stem-loop structures and the polyadenylate tract at the 3'-end of the genomic RNA are involved in the initiation of synthesis of minus strand RNA. Other viruses show higher order structures with likely functional roles in their life cycles. The need to conserve some RNA structures constitutes one of the limitations to nucleotide sequence variation in noncoding as well as in coding genomic regions (McKnight and Lemon, 1998): mutations that may disrupt functionally important RNA structures will be subjected to negative selection (Domingo and Holland, 1994).

The assumption that the ratio of nonsynonymous mutations (those that lead to amino acid substitutions; see also Chapter 6) to synonymous mutations (those that are silent regarding amino acid substitution) reflects the intensity of positive or negative Darwinian selection (Chapter 7) does not always apply to the evolution of RNA viruses. The reason is that the RNA genome itself may be part of the phenotype of the virus because of biological functions other than the protein-coding role. Silent base exchanges introduced into RNA genomes need not be neutral, as evidenced by selection pressure to produce revertants or pseudorevertants. Silent mutations in the replicase gene of phage Qβ were lethal even though the replicase produced showed normal activity (Mills et al, 1990). The same concept was dramatically illustrated in the reversion of a mutation associated with the temperature sensitive phenotype of poliovirus. In independent transfection experiments (generation of infectious virus following uptake of naked viral RNA into cells), in addition to the reversion of the phenotypically relevant, selected, nonsynonymous mutations, a number of synonymous replacements that had been introduced in the viral genome reverted as well (de la Torre et al, 1992). The synonymous replacements had been introduced into the poliovirus genome to provide restriction enzyme recognition sites to manipulate DNA copies of the genome, a common practice in molecular genetics, and it was completely unexpected that reversion to wild type would be selected frequently at these sites.

Computation of synonymous versus nonsynonymous mutations in comparisons of related genes often (but not always for viral genomes) results in a dominance of synonymous mutations. This was one of the arguments three decades ago to support the neutral theory of molecular evolution expounded by Motoo Kimura (Kimura, 1983). But there are many examples of short-term evolution of RNA viruses in cell culture which result in dominance of nonsynonymous versus synonymous mutations (Domingo and Holland, 1994). In these cases it cannot be excluded that in a coding region, silent changes may disrupt a function embodied in the RNA itself. Such silent changes may turn out to be more disruptive for the RNA structure than nonsynonymous changes disruptive for the functionality of the encoded protein. That is, a "silent" mutation need not be selectively neutral (Domingo and Holland, 1994).

Molecular Recognition of Nucleic Acids, and Selection for Nucleic Acid Structures—Compensatory Mutations

A crucial step in genome replication and gene expression is the recognition of key signals contained in nucleic acids. We mentioned previously that the structure of DNA allows the coding of any nucleotide sequence. This is not entirely correct. Arguments similar to those given in the preceding section on the phenotypic features contained in RNA genomes apply also to some extent to DNA. As in computer programs, there are restrictions in using key words that are reserved for a special purpose, e.g., a command. DNA does not only contain "extrinsic" information that can be expressed in proteins via RNA; it contains also intrinsic information: sequences that are recognized by regulatory proteins, enzymes or processors. For example, the long terminal repeat (LTR) of the immunodeficiency lentiviruses contains an array of signals for the recognition of host transcription factors or effectors encoded by the virus itself. When integrated as proviral DNA into the cellular DNA, these signals turn out to be essential for transcription of viral RNA as the first step to trigger the replication cycle of the virus (Coffin et al, 1997; Levy, 1998). The same and other signals in cellular DNA are needed for the regulated expression of cellular genes, and these signals (specific nucleotide sequences) must occupy defined positions in the DNA.

The recognition mechanisms for DNA-protein interactions follow some rules: different base-pairs have cognate residual hydrogen donor and acceptor groups in the large and the small grooves of the B-structure. Particularly, the contacts in the large groove are accessible to bulky protein structures like an α helix and allow an unequivocal identification of a base-pair. The base-pairs are not disrupted, and the basic structure of the DNA double helix is merely distorted by the interaction. Because of such recognition of base pairs, there are usually clearly defined consensus sequences for protein recognition on the DNA. Such sequences are not necessarily contiguous and may be interrupted by spacers of a random sequence, but with a defined chain length.

The recognition of RNA may be quite different. Often the protein or the enzyme makes contacts with hydrogen bond donors and acceptors that have a clearly defined distance. Because of the large number of possibilities of single strands to supply these contacts, it is usually difficult to identify a consensus sequence for RNA-protein interactions, even though the recognition may be highly specific. Examples of consensus sequences, e.g., the ribosome binding sites of prokaryotic messenger RNAs, or the termini of introns to be removed by splicing, usually involve processors that are ribonucleoproteins. In these examples the processors are ribosomes or spliceosomes, and the existence of consensus sequences are the consequences of Watson-Crick base-pairing between the processor and the target RNAs. Another revealing example is the IRES, a highly structured RNA region of about 450 residues comprising multiple stem-loops, bulges, and probably tertiary structures which have not yet been defined (Fig. 3.4). Several cellular proteins bind to viral IRES and contribute to efficient initiation of protein synthesis (Belsham and Sonenberg, 1996). When IRES from related viruses are compared, say among the picornaviruses (Chapter 2), differences in nucleotide sequences are abundant but structural conservation is striking. Often compensatory mutations have occurred which preserve a stem-loop structure attained with different primary sequences. In these cases the functional structure, and not the specific nucleotide sequence with which the structure is attained, is selected. Evolution has considerable room to act, but again negative selection will operate whenever mutations fail to produce specific shapes needed for IRES activity.

Most alterations of IRES elements that have been introduced by site-directed mutagenesis have resulted in reductions of translation-initiation activity. One exception was a mutation in the IRES of the picornavirus foot-and-mouth disease virus (FMDV) which resulted in a modest increase of activity. The mutation was a U → C transition at the base of loop 3, one of the major IRES structures composed of several subloops, stem-loops and bulges. Interestingly this mutation was selected in FMDV after prolonged persistence in cell culture, and probably contributed to the increased virulence of FMDV selected during persistence (Martínez-Salas et al, 1993). This modified IRES, derived from an evolving viral quasispecies, is now used in constructions for gene expression due to its enhanced activity in directing protein synthesis in a cap-independent manner. In protein-RNA interactions the situation may be even more complex than suggested in preceding paragraphs, since shapes as well as recognition of specific sequence are likely to play a role. Different types of protein-RNA interactions have been recognised for basic peptide chains and for nucleocapsid viral proteins (Patel, 1999).

Structure and Catalytic Properties of Proteins

Proteins are linear chains of 20 canonic alpha-amino acids connected by peptide bonds produced by the condensation of the amino group of one amino acid and the carboxyl residue of the preceding amino acid. Therefore each condensed chain, contains an amino and a carboxy terminus. The chain lengths range from short oligopeptides up to long chains built by a few thousands amino acids. The shape and the properties of a protein are a function of its amino acid sequence in the chain, and we thus see an important advantage of such polymers: if the sequence of the amino acid chain can be built reproducibly from the monomers, an enormous repertoire of different sequences and shapes can be created: of an average protein with a chain length of 400, 20^{400} (or 10^{520}) different molecules can be produced. These numbers are not only far beyond our imagination, they are also hyperastronomical: even if the whole universe were to be densely packed with a random library of proteins of chain length 400, the chance to find in it any specific chosen sequence would be vanishingly small. Together with the long range interactions of the monomers that cause the specific shape of a polypeptide with a certain sequence, one can build surfaces of almost any desired shape.

While nucleic acids have adequate properties to act as a repository of genetic information, proteins are ideally suited for catalysis and structure building. Amino acids provide a rich repertoire of polar, nonpolar, charged and electrically uncharged side chains, shapes and a wide range of hydrophylic and hydrophobic domains (Fig. 3.5). With the exception of proline, which is a planar imino acid, the different amino acids differ only in their side groups. The peptide structure forms the "backbone" of the protein structure from which the side groups protrude. The side groups may be hydrophilic or hydrophobic. As mentioned the hydrophobic part may form a solid core. While the backbone itself is polar, it may still be tolerated in a hydrophobic domain if the polar groups are saturated by intramolecular or intermolecular hydrogen bonding to other parts of the backbone. Two such structures of the backbone are predominant: the alpha helix and the beta sheet, the latter in a parallel or antiparallel conformation. Some amino acids have a positive or negative charge; they are normally hydrated and surrounded by counterions, but even these residues may occur in a hydrophobic domain, if two oppositely charged amino acids form a "salt bridge" that is additionally stabilized by hydrogen bonding. Even single water molecules may occur in hydrophobic pockets if their polar groups are saturated by hydrogen bonds to other groups.

In spite of the adequacy of nucleic acids for storage of information and of proteins for catalysis, these roles are not exclusive for each kind of polymer. RNAs can act as catalysts (Cech, 1986), and oligopeptides can act as templates for their own synthesis (Severin et al, 1997). It is not easy to propose specific models for the participation of these atypical capabilities of some RNAs and oligopeptides in the early evolution of life (Schuster and Stadler, 1999).

The structure of thousands of polypeptides have been investigated by X-ray crystallography or by nuclear magnetic resonance (NMR) studies, and the coordinates of their atoms are deposited in data banks. This is remarkable, since the number of degrees of freedom in folding a chain are of course at least as hyperastronomical as the number of alternatives in combining the monomers. Nevertheless, proteins fold rapidly during synthesis into a defined form and generally do not produce mixtures of different structures. However, X-ray crystallography studies have shown that the coordinates of some atoms, in particular those of hydrophobic cores, are precisely fixed while those of others, preferentially those in contact with the solvent, may vary within a rather limited radius. Some portions of the chain are folded into a rather rock-solid "domain" while other parts of the sequence are flexible hinges which allow changes in the conformation (for review of protein structure see Creighton, 1993 and Kyte, 1995).

Protein structures can become disrupted in a process called "denaturation", e.g., by heat or by nonpolar solvents. The energy gain produced by the intramolecular bonding has to be overcome for that reaction, thus making it a strongly endothermic one, but there is a large gain in entropy since the denatured protein is no longer in a defined ordered structure. It is surprising that for some small proteins the correct folding can be restored by a slow "renaturation". Therefore, the correct form is also the thermodynamically most stable one. Energy calculations, on the other hand, indicate there is not a single predominant energy minimum, but

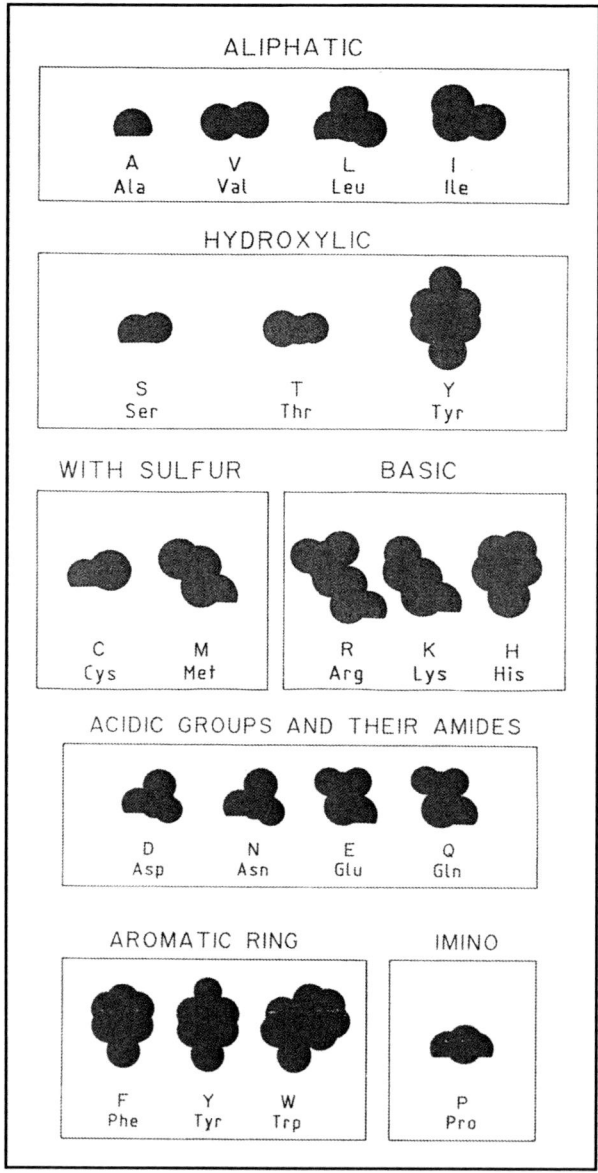

Fig. 3.5. Classification of amino acids with a side chain other than hydrogen. In each box the main contour feature of each related side-chain is indicated. Amino acid contours have been drawn based on the program Chem 3D Plus™. Orientation is chosen so as to produce the maximum silhouette after erasing hydrogens and drawing all other atoms in black. Glycine (Gly, G) would classify as aliphatic, but it has been excluded because its side chain is an hydrogen atom. The outlines have been drawn based on Fig. 4-16 of Kyte (1995). (With permission from Garland Publishing, Inc. New York and London).

rather several isolated local minima of similar energy. How can then the molecule find its proper folding? Why does the structure of some molecules not get trapped in local minima? Obviously, if only the thermodynamic energy contributions played a role, it would be almost impossible for a protein to find its energetically most-favored structure in a finite time. Renaturation must thus be accelerated by kinetic guidance: at the very beginning, a few particularly strong interactions are formed that seriously reduce the number of degrees of freedom, and lead to a rather defined renaturation pathway. Stabilization is also facilitated by cross-linking with S-S bridges between cysteine residues in the polypeptide chain or by other specific interactions (Kyte, 1995). It is conceivable that evolution has gradually selected proteins also for their ability to acquire efficiently their defined, functional structures. Denaturation requirements vary strongly from protein to protein; even sequentially related proteins usually denature at different temperatures. Proteins from organisms adapted to live at high temperatures or "thermophiles" are particularly adapted to resist denaturation at these conditions. They are the source of thermostable DNA polymerases which are routinely used for amplification of genetic material in vitro since they tolerate temperatures needed for sequential annealing and melting of primer-template pairs (Chapter 8). In viruses, a lower probability of denaturation (when the latter implies functional losses) of virion proteins must be a trait selected in their long-term evolution, as viruses must survive as free infectious particles throughout many transmission events. Indeed, the resistance of virus to denaturing agents depends on their infection mode: the enterovirus group of picornaviruses is much more resistant to acid than the rhinoviruses, because enteroviruses have been selected to survive the passage through the stomach.

Quaternary Structures

Long-range interactions to create structures and hydrophobic cores are not restricted to atoms within the same polypeptide chain. Often such interactions make stable quaternary complexes between a number of molecules of the same or of other sequences. Many enzymes contain several polypeptide subunits. The protein capsids of viruses are defined quaternary structures with many subunits. Many nonenveloped viruses are in fact a single molecular complex with a defined quaternary structure in which each atom has a rather strongly defined coordinate. Many of these viruses have been crystallized and their structures determined by x-ray crystallography.

Most biochemical reactions are catalyzed by proteins: the enzymes. A catalyst is defined as a component that increases the rate of a chemical reaction without being consumed. In the simplest formulation of Michaelis and Menten, an enzyme binds reversibly to a substrate. The enzyme-substrate complex is then transformed into a product, releasing the free enzyme. Many enzymes catalyze more complex reactions than just the transformation of a substrate molecule into a product. Furthermore, a number of proteins exist as multifunctional enzymes. This occurs in the case of several enzymes involved in nucleic acid replication (described in following sections). Enzymes can accelerate a chemical reaction by factors of 10^6 up to 10^{10}. In many cases, in particular in reactions involved in expression of genetic information, the acceleration factor cannot even be determined because the reaction rate in the absence of the catalyst is immeasurably slow. The increase of reaction rate involves several mechanisms. The collision of two appropriately-aligned reactants can be favored by their simultaneous binding to aligned enzyme sites, thus increasing their effective concentrations. The binding energy is used to distort the conformation of a reactant towards its transition state structure. Groups of one reactant may be transferred to groups of the other reactant. Acid-base catalysis is a particular common example: less energy is needed to transfer protons to side chains of enzyme amino acids such as histidine than to water. Sometimes the distinction between a catalyst and a reactant is not obvious. A coenzyme may sometimes work as a coreactant, but it may be recycled by a second, independent reaction. The entire metabolism of a cell may be viewed as a network of many such interconnected reaction cycles producing functional "pathways".

Evolution has probably shaped catalytic reactions to maintain/express pathways that are very slow in the absence of biocatalysts. This is, of course, achieved by mutation/selection of nucleic acid sequences in genomes. Spontaneous reactions that could interfere with the normal development of biological activities had to be suppressed. This may be why biological reactions involving phosphate are common, since a high energy of activation is needed to obtain a trigonal bipyramidal transition state (Westheimer, 1987). In a few cases, slow modifications of the genome occur without biological catalysis. An example is methylation of bases by S-adenosylmethionine. The modification may be reversed by an enzyme-catalyzed reaction in DNA, and possible mutagenic consequences for RNA viruses must be tolerated. Enzymes can obviously be modified in their catalytic constants and specificity. In nature, enzymes encoded by cells and viruses are subjected to processes of mutation, competition and selection. "Test tube" modification of enzymes to derive enzymes with new catalytic activities is an important part of a growing field of research termed evolutionary biotechnology (Chapter 8).

Replication Enzymes

Essential to all modern life forms is the existence of a protein machinery capable of copying nucleic acid chains into mutually complementary sequences. For single stranded viral RNA genomes, the complementary RNA is a replicative intermediate which serves in turn as the template to synthesize new progeny RNA genomes. Nucleic acid synthesis (replication) is the means for maintaining the genetic information encoded in the nucleotide sequence of any genome and for spread of progeny copies with the same (or very similar) information. The monomeric precursors are nucleoside 5'-triphosphates from which one nucleotide complementary to the nucleotide in the template is transferred to the 3'-terminus of the growing chain, releasing free pyrophosphate as the leaving group. Nucleic acid synthesis in the growing chain proceeds in the 5'-3' direction, as a result of copying the template from the 3' to the 5' direction. Several types of enzyme designs have been shaped by viral evolution to achieve template recognition and polymerization of monomeric nucleoside triphosphates (phosphodiester bond formation). Two major categories have been recognized: enzymes made solely of proteins encoded by the viral genome, and enzymes made of virus-coded proteins together with proteins encoded by the host. The best characterized enzymes of the first category are the retroviral reverse transcriptases (RT) (Skalka and Goff, 1993; Coffin et al, 1997), in particular the RT of human immunodeficiency virus type 1 (HIV-1), an enzyme that has been the object of intense investigation over the last decade, in part due to the pressing need to design specific inhibitors of RT as anti-HIV-1 agents (Levy, 1998; Jonckheere et al, 2000). This enzyme is a dimer composed of a catalytic subunit of 66 kDa and a 51 kDa subunit which is derived from the large 66 kDa subunit by proteolytic cleavage. It is a multifunctional enzyme in that it is capable of copying both RNA and DNA into DNA, and it contains also a ribonuclease H (RNase H) domain at the carboxy-side of the 66 kDa subunit. The RT of HIV-1 has been crystallized in several laboratories as a complex with a variety of ligands, and this has allowed structural studies with the different domains involved in nucleic acid binding and polymerization functions. A combination of biochemical and structural approaches is currently serving to define the molecular basis of substrate recognition and of the copying fidelity of RT (Skalka and Goff, 1993). It appears from recent results that fidelity of substrate recognition is not a fixed property inherent to the catalytic activity of the enzyme, but rather, one which can be modified by subtle structural alterations around the catalytic site (see section on Fidelity of RNA and DNA polymerases, later in this Chapter).

A well known example of a viral polymerase composed of virus-coded as well as host proteins is the Qβ replicase, which can be purified from bacteriophage Qβ-infected *Escherichia coli* cells. This enzyme provided the basis for the first in vitro RNA replication system capable of catalyzing the test tube synthesis of infectious viral RNA from nucleotide substrates under the required ionic conditions. This work was initiated by Sol Spiegelman and associates, and then followed by the groups of C. Weissmann, D.R. Mills, F.R. Kramer, C. Biebricher, and others. Qβ replicase can amplify infectious Qβ RNA in vitro via a complementary strand, and it can also replicate small variant RNAs of a few hundred nucleotides in length. Studies with small variant RNAs have been instrumental in defining the molecular mechanisms of replication and in the development of concepts of high mutability, competition and selection as essential features of RNA genetics (Biebricher and Gardiner, 1997) (see also Chapters 6, 7 and 8). The Qβ replicase which is described in detail in following sections is composed of a virus-coded subunit and three host subunits. It provides the ultimate example of molecular parasitism because host proteins which perform important functions for protein synthesis in the uninfected *Escherichia coli* cell are coopted, sequestered and assembled together with a viral protein to ensure multiplication of the invader phage. Other virus groups use an amazing variety of protein resources to ensure replication of their genetic material (Coen, 1996). For example, multiple proteins, including a DNA-dependent DNA polymerase, are encoded in the highly complex genomes of herpes simplex virus, the related varicella-zoster virus, and cytomegalovirus. In these cases several viral-coded protein are involved in viral DNA replication.

In spite of the great diversity of biological approaches which have evolved to achieve replication of DNA and RNA (Table 3.1), the comparison of the three-dimensional structures of cellular and viral DNA polymerases (for example, the DNA polymerase of *Escherichia coli*, the retroviral RT, and the poliovirus RNA-dependent RNA polymerase) show a number of common features (Sousa, 1996; Hansen et al, 1997). One of them is a conserved general shape with nearly identical residues and domains involved in catalysis. Such structural comparisons have also revealed an important feature of RT and viral RNA-dependent RNA polymerases: the absence of a 3'-5' exonucleolytic activity needed for proofreading of misinserted nucleotides, an activity present as a distinct domain in several cellular and viral DNA polymerases, such as those of bacteriophage T4 or herpes simplex virus (Sousa, 1996; Hall et al, 1996). The lack of proofreading and repair activities contributes to the elevated error rates of viral replicases and reverse transcriptases since when such error-correcting mechanisms are suppressed in cellular DNA polymerases, these enzymes reach error levels typical of RNA viruses. Thus the absence of proofreading-repair activities is one of the biochemical bases of error-prone replication, a hallmark of RNA genetics. Structural studies with nucleic acid polymerases have also provided evidence for a modular origin of such enzymes, with distinct functions associated with particular structural domains (Sousa, 1996).

DNA-Dependent DNA Polymerases

The DNA-dependent DNA polymerases are key enzymes in three important genetic processes: DNA replication, DNA repair, and DNA recombination. They are usually specialized for one specific purpose. Nevertheless, their basic characteristics are quite similar. They use as precursors the magnesium complexes of nucleoside triphosphates, and can only fill gaps in the 5' → 3' direction. Chromosomal DNA replication starts at one or several unique sites termed the origins of replication. DNA synthesis may proceed in one or both directions from the origin. Usually one strand is synthesized continuously and the other discontinuously. Many replication proteins which can be viewed as a multienzyme complex assemble at the origin of replication or at the replication fork to participate in nucleotide polymerization. For example, the DNA polymerase III holoenzyme of *Escherichia coli* is made of at least ten polypeptides,

Table 3.1. Properties of DNA and RNA polymerases

Enzyme	Template	Product	Primer-dependent?	Sequence-specific?	Role in Vivo	Source
DNA polymerase	ss-DNA cs	ds-DNA	Yes	– –	DNA repair and replication transcription	all organisms
RNA polymerase	ds-DNA rc	ss-RNA	No	+ –	transcription	all organisms
Reverse transcriptase	ss-RNA ss-DNA cs	RNA:DNA ds-DNA	Yes Yes	–	virus replication	retroviruses, retroelements
RNA replicase	ss-RNA rc	ss-RNA	No	++	virus infected replication	levivirus bacteria
Chemical	ss-RNA cs	ds-RNA	Partial	–	none	artificial

Notes: cs-template consumed; rc-template recycled; ss-single-stranded; ds-double-stranded. In template-instructed chemical polynucleotide synthesis experiments (reviewed by Orgel, 1987), the primer requirement and the sequence specificity were found to vary depending on the templates and the conditions used. The + – notation for RNA polymerase indicates the sequence-specific initiation at a promoter site followed by nonspecific elongation; the + + notation for RNA replicase indicates sequence specificity for both initiation and elongation, and the – – notation for DNA polymerase indicates lack of sequence specificity for both.

and additional proteins participate in the different steps of initiation, elongation and termination of DNA synthesis (Kornberg and Baker, 1992; DePamphilis, 1996). Among others, single-stranded DNA binding proteins prevent base pairing and maintain DNA chains in an extended conformation at the replication fork, DNA topoisomerases release torsional stress, and DNA ligase links discontinuous DNA fragments. DNA helicase mediates the unwinding of the double helix to provide single stranded template regions at the replication fork. Unwinding requires much energy for the rotation of the double helix around its axis. The 3' → 5' proofreading exonuclease activity is an activity associated with the polymerase III holoenzyme.

Contrary to RNA synthesis, DNA synthesis requires a preexisting primer for initiation. Several types of primers are used (Salas et al, 1996): terminal 3'-OH ends of DNA, the 3'-OH of t-RNA molecules (such as in initiation of reverse transcription of retroviral RNA), nascent transcribed RNA chains, and a deoxyribonucleoside monophosphate covalently linked to a Ser, Thr or Tyr residue of a protein. This is the case for φ29, the adenoviruses or hepatitis B virus, among others. For bacterial DNA synthesis, a primase synthesizes a short RNA transcript which serves as primer for deoxynucleotide polymerization. The lagging strand—which elongates away from the direction in which the replication fork is proceeding—requires synthesis of multiple RNA primers for the synthesis of short DNA fragments termed the Okazaki fragments. In eukaryotic cells, DNA replication is triggered by multiple signals with mechanisms which are not well understood. DNA replication is also one of the processes involved in repair of DNA damage. For example, following the excision of damaged DNA, the eukaryotic DNA polymerases δ and σ may fill the gap by synthesizing DNA. Structural and biochemical studies with a number of DNA-dependent and RNA-dependent DNA polymerases (reverse transcriptases) are currently contributing to define mechanisms of nucleotide incorporation and copying fidelity (Steitz, 1993, 1999).

Because of the complexity of DNA replication it is perhaps not surprising that until very recently it has not been possible to use an in vitro DNA replication system able to attain large amplification of specific DNA segments. This now has been achieved with the "polymerase chain reaction" (PCR) using thermostable DNA polymerases and specific sets of primers.

RNA Replication

RNA-dependent RNA polymerase activities have been detected in some plant cells, but are apparently absent in animal or bacterial cells. These activities are typically associated with the replication of RNA viruses (strategy 1 of Fig. 2.3). An in vitro amplification of template RNA has been described only for the plant virus cucumber mosaic virus (Hayes and Buck, 1990), and for the *leviviridae*, RNA bacteriophages infecting several enterobacterial species (Zinder, 1975). Because of its stability, the replicase of coliphage Qβ has been extensively studied and used as a model system in studies of the mechanism of RNA replication and in vitro evolution (Chapter 5).

When a single Qβ RNA strands invades a host cell, it encounters about 10^3 strands of mRNA, 10^4 copies of ribosomal RNA (rRNA) and 10^5 molecules of transfer RNA (tRNA). A broad-range RNA replicase in the cell would thus lead to a disorganized RNA amplification with little, if any, production of progeny virus, a condition sine qua non for being a virus. All RNA replicases that have been studied have been found to be highly specific for their cognate viral RNA, ignoring largely the RNA from the host.

Replication of Qβ RNA starts with the production of a complementary minus strand (Weissmann, 1974), often isolated from infected cells to be associated with the viral RNA strand to form an antiparallel Watson-Crick type double helix. Careful studies have revealed that the double strand is an isolation artefact: the minus strand is separated during the synthesis from its template and liberated in single-stranded from. The minus strand serves as a template for synthesis of full-strand plus strand.

In contrast to the sophisticated DNA replication apparatus, the RNA replication system is very simple: a single enzyme catalyses, together with a host factor is able to perform, all replication steps. The replicase has been purified to homogeneity (Kamen, 1970; Kondo et al, 1970). It contains four subunits, only one encoded by the virus while the others were identified as the ribosomal protein S_1 and the protein biosynthesis elongation factors (EF)Tu and (EF)Ts. The viral subunit seems to provide the active center for phosphodiester bond formation (see also Fig. 2.5 in Chapter 2).

A number of short-chained RNA species that are efficiently amplified by Qβ replicase have been isolated and characterized. Because of their high template efficiency and their easy handling, these RNA species were used to investigate the replication mechanism and to study in vitro evolution of RNA (Chapter 5).

Reverse Transcriptase

Reverse transcriptase (RT) or RNA-dependent DNA polymerase is encoded in the genomic RNA of all retroviruses (Skalka and Goff, 1993; Arts and Le Grice, 1998), in a number of cellular retroelements, and also in the hepadnavirus *pol* gene. There is considerable diversity in the subunit composition of RT from different retroviruses and from other origins. The murine leukemia virus RT consists of a single protein subunit which includes both a polymerase and an associated RNase H activity. RNase H degrades RNA when present in an RNA-DNA hybrid. In contrast, both the avian sarcoma-leukosis viruses (ASLV) and the human immunodeficiency viruses (HIV) assemble an RT made of two subunits of different size the smaller one being a part of the larger one. In the case of ASLV the large subunit includes also an integrase activity which in the other retrovirus groups is present in a separate protein. The integrase is an enzyme that mediates the integration of a DNA copy of the retroviral genome into the chromosomal DNA of the host cell, an essential step in retroviral replication (Fig. 2.9).

The organization of the *pol* genes which encode these different enzymes illustrates not only how different modules have been combined, but also how these modules relate to viral replication strategies. The *pol* region of HBV does not encode a protease function but it encodes the terminal protein required for initiation of viral DNA replication. It does not encode an integrase-nuclease function either, and this could be anticipated from the fact that integration into the cellular DNA is not part of the standard replication cycle of this virus. Both retroviruses and hepadnaviruses have been selected to include some viral nucleic acid reservoir in a form that may trigger viral expansions. Reservoirs consist of integrated DNA in the case of HIV-1, and of covalently closed circular DNA molecules (maintained in the cell nuclei in episomal form, without integration) in the case of HBV. These viral DNA reservoirs contribute to viral persistence in the host, and represent a means for potential new particles to elude defensive responses of the host, or to hide from damaging environmental alterations such as the presence of antiviral agents. Alternative molecular strategies leading to latency, chronicity and persistence are displayed by other viruses (Chapter 6).

RT is found in, and exerts its copying activity within, the nucleocapsid of retroviral particles which include two copies of genomic RNA. It initiates minus strand DNA synthesis on a t-RNA primer annealed near the 5'-terminus of one genomic RNA molecule. After copying into DNA, the copied RNA is degraded by the RNAse H associated with the RT. Then the enzyme with the nascent DNA is transferred to the homologous region at the 3' end of the second genomic RNA (first strand transfer). The minus strand DNA is completed and the RNA template is degraded by the RNAse H. A RNase H-resistant polypurine tract serves as primer for the synthesis of positive strand DNA. Following the RNase H-mediated cleavage of the primer t-RNA, a second strand transfer permits completion of the double stranded DNA. This can then integrate into the host DNA and replicate as provirus with the cellular DNA. The productive replication cycle is initiated by transcription of the integrated provirus catalyzed by the cellular DNA-dependent RNA polymerase II and mediated by a number of cellular transcription factors and viral regulatory elements.

One important observation made in recent years from the analysis of RT of HIV-1 from patients undergoing treatment with RT inhibitors is the surprising tolerance of this enzyme for amino acid replacements around the catalytic domain while retaining function (see Chapter 8 for specific examples and references). This has been one of the most disappointing findings for the treatment of AIDS. Furthermore, contrary to the initial expectations of a more limited tolerance to substitutions in the HIV-1 protease due to its smaller size, mutations that render the protease insensitive to inhibitors also occur with high frequencies. These observations have forced modification of the belief that viral enzymes would tolerate far fewer amino acid substitutions than structural proteins such as capsid or surface proteins of viral particles. However, in a comparison of the mutation frequencies of *gag*, *env* and *pol* of HIV-1, mutation frequencies in *env* were found to be at most two- to three-fold larger than in *pol* (Quiñones-Mateu et al, 1996) (see also Chapter 8). Even considering selective constraints on highly essential domains of catalytic viral proteins, viruses may find ways (through specific second-site mutations, constellations of compensatory mutations, etc.) to overcome such constraints. This is why antiviral strategies based solely on inhibition of viral functions are likely to fail unless resistant mutations entail viral extinction. After all, viruses have been selected to overcome such inhibitory activities which occur naturally in the hosts that they parasitize. Unavoidably, an equilibrium had to be attained that resulted in survival of both the viruses and their hosts. This is why viruses still exist, and some of their hosts can still write about them.

Fidelity of RNA and DNA Polymerases and the Survival of Defective Genomes

The absence of a 3'-5' proofreading exonuclease activity and of other post-replicative repair mechanisms which operate on double stranded DNA (Friedberg et al 1995) is believed to be one of the major influences underlying the error-prone nature of RNA genome replication (see also Chapter 4). Cellular DNA-dependent RNA synthesis (transcription) is also inaccurate since cellular RNA polymerases appear to be deficient in error-correcting mechanisms. However, a 3' → 5' nuclease activity of human RNA polymerase II shows proofreading in vitro (Thomas et al, 1998). It is not clear what might be the role of such correcting activity in vivo. In transcription the reading accuracy of nucleotides as dictated by their chemical nature may suffice, since errors are neither propagated to other RNA molecules nor transmitted to offspring of the cell. Similar arguments apply to translation errors. What is needed is a minimal amount (a very high percentage) of functional RNA and protein molecules compatible with metabolic activity. Presumably, there is no strong selection favoring error-correcting mechanisms during transcription or translation, as there is for DNA replication. When RNA itself is replicated, however, error propagation has serious consequences. RNA viruses have to live with this limited replication accuracy, and they are also able to exploit it for adaptation. The amount of information that RNA viruses can transmit, and thus their chain lengths, are constrained by replication errors. Once the constraint is accommodated, errors serve for adaptation to changing host environments.

Can fidelity of viral replicases be modified? Early work on genetics of bacteriophage T4 suggested that some phage polymerase mutations increased while others decreased mutation frequencies, and the proofreading 3'-5' exonuclease activity contributed to copying accuracy (Drake et al, 1969; Muzyczka et al, 1972; Reha-Krantz, 1994; Goodman and Fygenson, 1998). Genetic studies with RNA viruses suggested that mutation rates can vary among individual viral clones (Pringle et al, 1981; Suarez et al, 1992). It is also known that different enzymes, ionic environments and template sequences may lead to considerable variations in copying accuracies in vitro (Ricchetti and Buc, 1990; Goodman and Fygenson, 1998). Recent results of site-directed mutagenesis of the RT-coding regions have indicated that some amino acid substitutions around the catalytic domain of this enzyme may enhance its copying fidelity, while other substitutions may lower its fidelity (Wainberg et al, 1996; Martín-Hernández et al, 1996). Thus, there is now evidence that the fidelity of viral polymerases may be amenable to modification, even in the absence of a 3'-5' exonuclease activity which could modulate overall accuracy. The recent results imply that copying fidelity of viral replicases or retrotranscriptase may be modified by externally applied influences (drugs that may distort the catalytic domains, or the like). This is not without consequence in considering alternative antiviral strategies that avoid viral inhibition as their central mode of action (Chapter 8).

It must be emphasized that because of multiple environmental factors that can affect mutation rates in vitro, it is difficult to compare results of different fidelity assays, and even similar assays performed in different laboratories. Not surprisingly, mutant RTs displaying indistinguishable misincorporation fidelity using a specific test, produced different overall mutational spectra in progeny nucleic acids (Drosopoulos and Prasad, 1998).

In contrast to cellular genomes, viral genomes are not confined to different compartments after each duplication step. Late in an infection cycle many copies of the RNA genome are present together with their expression products. In this case, selection pressure acts on the whole genome distribution, since some mutants unable to establish a successful infection on their own can still be amplified in a cell. Unfit (or even completely defective) mutants can survive by complementation, that is, by the supply of biologically active proteins by a competent (often called helper) infectious virus. Debilitated or defective mutants can survive several infection cycles if cells are infected with high numbers of particles per cell to ensure complementation (see section on Defective Viral Genomes in Chapter 2). Recloning and selection for the full information of the individual genome is observed only when a single virus invades a new cell. This "swamping" effect also favors genetic variability and allows "hitch-hiking" of newly introduced mutations. Thus, limited replication fidelity can cause a substantial proportion of viral genomes to be defective because of highly deleterious genetic lesions: point mutations, insertions, deletions, homopolymeric extensions, rearrangements, etc. Quasispecies act as a unit of selection and complementation among viruses of unequal fitness may also occur.

References

1. Arora R, Priano C, Jacobson AB, Mills DR. Cis-acting elements within an RNA coliphage genome: Fold as you please, but fold you must!!! J Mol Biol 1996; 258:433-46.
2. Arts EJ, Le Grice SFJ. Interaction of retroviral reverse transcriptase with template-primer duplexes during replication. Progress in Nucl Acid Res 1998; 58:339-93.
3. Belsham GJ, Sonenberg N. RNA-protein interactions in regulation of picornavirus RNA translation. Microbiol Rev 1996; 60:499-511.
4. Biebricher CK, Gardiner WC. Molecular evolution of RNA in vitro. Biophysical Chemistry 1997; 66:179-92.
5. Cech TR. RNA as an enzyme. Sci Am 1986; 255:76-84.
6. Coen DM. Viral DNA polymerases. In: DePamphilis ML, ed. DNA replication in eukaryotic cells. New York: Cold Spring Harbor Laboratory Press 1996:495-523.
7. Coffin J, Hughes SH, Varmus HE, eds. Retroviruses. New York: Cold Spring Harbor Laboratory Press, 1997.
8. Creighton TE. Proteins. Structures and molecular properties. New York: W.H. Freeman and Company 1993.
9. de la Torre JC, Giachetti C, Semler BL et al. High frequency of single-base transitions and extreme frequency of precise multiple-base reversion mutations in poliovirus. Proc Natl Acad Sci USA 1992; 89:2531-2535.
10. Deiman BALM, Pleij CWA. Pseudoknots: A vital feature in viral RNA. Seminars in Virology 1997; 8:166-175.
11. DePamphilis ML. DNA replication in eukaryotic cells. New York: Cold Spring Harbor Laboratory Press, 1996.
12. Domingo E, Holland JJ. Mutation rates and rapid evolution of RNA viruses. In: Morse SS, ed. The Evolutionary Biology of Viruses. New York: Raven Press 1994:161-84.
13. Drake JW, Allen EF, Forsberg SA et al. Genetic control of mutation rates in bacteriophage T4. Nature 1969; 221:1128-32.
14. Drosopoulos WC, Prasad VR. Increased misincorporation fidelity observed for nucleoside analog resistance mutations M184V and E89G in human immunodeficiency virus type 1 reverse transcriptase does not correlate with the overall error rate measured in vitro. J Virol 1998; 72:4224-30.
15. Friedberg EC, Walker GC, Siede W. DNA repair and mutagenesis. Washington DC: ASM Press, 1995.
16. Goodman MF, Fygenson DK. DNA polymerase fidelity: From genetics toward a biochemical understanding. Genetics 1998; 148:1475-82.
17. Hall JD, Orth KL, Claus-Walker D. Evidence that the nuclease activities associated with the herpes simplex type 1 DNA polymerase are due to the 3'-5' exonuclease. J Virol 1996; 70:4816-18.
18. Hansen JL, Long AM, Schultz SC. RNA polymerase of poliovirus. Structure 1997; 5:1109-22.
19. Hayes RJ, Buck KW. Complete replication of a eukaryotic virus RNA in vitro by a purified RNA-dependent RNA polymerase. Cell 1990; 63:363-8.
20. Jonckheere H, Ammé J, De Clercq E. The HIV-1 reverse transcription (RT) process as target for RT inhibitors. Med Res Rev 2000; 20:129-54.
21. Kamen IR. Characterization of the subunits of Qβ replicase. Nature 1970; 228:527-533.
22. Kimura M. The neutral theory of molecular evolution. Cambridge: Cambridge University Press; 1983.
23. Kyte J. Structure in protein chemistry. New York and London: Garland Publishing, Inc. 1995.

24. Kondo M, Gallerani R, Weissmann C. Subunit structure of Qβ replicase. Nature 1970; 228: 525-527.
25. Koop BF, Rowan L, Chen WQ et al. Sequence length and error analysis of sequenase and automated *Taq* cycle sequencing methods. Biotechniques 1993; 14:442-447.
26. Kornberg A, Baker T. DNA replication. New York: WH Freeman, 1992.
27. Levy JA. HIV and the pathogenesis of AIDS. Washington D.C.: ASM Press, 1998.
28. Lundberg KS, Shoemaker DD, Adams MW et al. High-fidelity amplification using a thermostable DNA polymerase isolated from *Pyrococcus furiosus*. Gene 1991; 108:1-6.
29. Martín-Hernández AM, Domingo E, Menéndez-Arias L. Human immunodeficiency virus type 1 reverse transcriptase: role of Tyr-115 in deoxynucleotide binding and misinsertion fidelity of DNA synthesis. EMBO J 1996; 15:4434-42.
30. Martínez-Salas E, Sáiz JC, Dávila M et al. A single nucleotide substitution in the internal ribosome entry site of foot-and-mouth disease virus leads to enhanced cap-independent translation in vivo. J Virol 1993; 67:3748-55.
31. McKnight KL, Lemon SM. The rhinovirus type 14 genome contains an internally located RNA structure that is required for viral replication. RNA 1998; 4:1569-84.
32. Mills DR, Prisno C, Merz PA et al. Qβ replicase: Mapping the functional domains of an RNA-dependent RNA polymerase. J Mol Biol 1990; 205, 751-764.
33. Muzyczka N, Poland RL, Bessman MJ. Studies on the biochemical basis of mutation. I. A comparison of the deoxyribonucleic acid polymerases of mutator, antimutator, and wild type strains of bacteriophage T4. J Biol Chem 1972; 247:7116-22.
34. Nowakowski J, Tinoco I. RNA structure and stability. Seminars in Virology 1997; 8:153-65.
35. Olsthoorn RCL, van Duin J. Random removal of inserts from an RNA genome: Selection against single stranded RNA. J Virol 1996; 70:729-36.
36. Orgel LE. Evolution of the genetic apparatus: A review. Cold Spring Harbor Symp Quart Biol 1987; 52:9-16.
37. Patel DJ. Adaptive recognition in RNA complexes with peptides and protein modules. Curr Op Struct Biol 1999; 9:74-87.
38. Pringle CR, Devine V, Wilkie M et al. Enhanced mutability associated with a temperature sensitive mutant of vesicular stomatitis virus. J Virol 1981; 39:377-89.
39. Quiñones-Mateu ME, Holguín A, Dopazo J et al. Point mutant frequencies in the *pol* gene of human immunodeficiency virus type 1 are two- to three-fold lower than those of *env*. AIDS Res Hum Retroviruses 1996; 12:1117-28.
40. Reha-Krantz L. Genetic dissection of T4 DNA polymerase structurefunction relationships. In: Karam JD ed. Molecular biology of bacteriophage T4. Washington DC: American Society for Microbiology 1994: 307-12.
41. Ricchetti M, Buc H. Reverse transcriptases and genomic variability: the accuracy of DNA replication is enzyme specific and sequence dependent. EMBO J 1990; 9: 1583-93.
42. Saenger W. Principles of nucleic acid structure. New York: Springer-Verlag, 1984.
43. Saiki RK, Gelfand DH, Stoffel S et al. Primer-directed enzymatic amplification of DNA with a thermostable DNA polymerase. Science 1988; 239:487-491.
44. Salas M, Miller JT, Leis J et al. Mechanisms for priming DNA synthesis. In: DePamphilis ML, ed. DNA replication in eukaryotic cells. New York: Cold Spring Harbor Laboratory Press 1996: 131-75.
45. Schuster P, Stadler PF. Nature and evolution of early replicons. In: Domingo E, Webster RG, Holland JJ, eds. Origin and evolution of Viruses. San Diego: Academic Press; 1999:1-24.
46. Severin K, Lee DH, Granja JR et al. Peptide self-replication via template directed ligation. Chemistry 1997; 3:1017-24.
47. Skalka AM, Goff SP. Reverse Transcriptase. New York: Cold Spring Harbor Laboratory Press, 1993.
48. Sousa R. Structural and mechanistic relationships between nucleic acid polymerases. Trends Biochem Sci 1996; 21:186-90.
49. Steitz TA. Similarities and differences between RNA and DNA recognition by proteins. In: Gesteland RF, Atkins JF, ed. The RNA world. New York: Cold Spring Harbor Laboratory Press 1993; 219-37.
50. Steitz TA. DNA polymerases: Structural diversity and common mechanisms. J Biol Chem 1999; 274:17395-17398.

51. Suarez P, Valcarcel J, Ortín J. Heterogeneity of the mutation rates of influenza A viruses: Isolation of mutator mutants. J Virol 1992; 66:2491-94.
52. Thomas MJ, Platas AA, Hawley DK. Transcriptional fidelity and proofreading by RNA polymerase II. Cell 1998; 93:627-37.
53. Wainberg MA, Drosopoulos WC, Salomon H et al. Enhanced fidelity of 3TC-selected mutant HIV-1 reverse transcriptase. Science 1996; 271:1282-85.
54. Watson JD, Crick FHC. A structure for deoxyribose nucleic acid. Nature 1953; 171:7373-78.
55. Weissmann C. The making of a phage. FEBS Letters 1974; 40:S10-S12.
56. Westheimer FH. Why nature chose phosphates. Science 1987; 235:1173-78.
57. Wyatt JR, Tinoco I. RNA structural elements and RNA function. In: Gesteland RF, Atkins JF, eds. The RNA world. New York: Cold Spring Harbor Laboratory Press 1993:465-96.
58. Xiang WK, Paul AV, Wimmer E. RNA signals in entero-and rhinovirus genome replication. Seminars in Virology 1997; 8:256-73.
59. Zinder ND, ed. RNA phages. New York: Cold Spring Harbor Laboratory, 1975.

CHAPTER 4

Quantitative Molecular Evolution
Autocatalytic Growth

Darwinian evolution, instrumental in the qualitative description of many phenomena in Biology, can and should also be formulated quantitatively. Darwin himself cited the population growth law described by Thomas R. Malthus in *An Essay on the Principle of Population Growth as it Affects the Future Improvement on Society*: Unrestricted growth behaves as a geometric series; the population increases exponentially. The potential for exponential growth is indeed a prerequisite for Darwinian evolution, independent of the reproduction mechanism, for mother organisms reproduce at a rate proportional to their number. When describing Mendelian populations and sexual reproduction, only the number of females need be taken into account, assuming that enough males are present to fertilize the females.

Quantitative measurements are particularly easy for microorganisms or viruses. Let us consider the growth of a bacterial culture in a nutrient broth. Bacteria have a simple reproduction mechanism: The cell volume increases by taking up and metabolizing nutrients until the cell eventually divides into two identical daughter cells; the daughter cells mature to mother cells and then divide again.

The simplest measurement is field observation of duplication times (Fig. 4.1): We observe an immobile bacterium under the microscope. After a certain time period, the bacterium has divided and two daughter cells are present; after a further time period another division occurs. The elapsed times τ between two consecutive divisions are roughly equal, the recorded values scattering around the average value. By the third or fourth division all synchronization is lost: The bacteria no longer divide simultaneously. On the other hand, if we measure a larger bacterial population by plating, i.e., we distribute an aliquot of the solution on the surface of a solid nutrient medium and count the number of colonies that have developed after incubating the plate, we find that the duplication times τ are quite reproducible: After n divisions, the population has increased according to $N(n) = N(0) \cdot 2n$, or after the time t:

$$N(t) = N(0) \cdot 2^{t/\tau} \qquad (4.1)$$

If the starting population were perfectly homogeneous and the growth perfectly synchronous, the law would require integer divisions of t/τ, giving a staircase growth curve. In the experiment, the synchrony is lost after a few divisions and continuous exponential growth is observed; a stable age distribution has been formed wherein all age classes show coherent exponential growth.

The division time τ depends on a number of environmental influences such as nutrient composition, oxygen tension or temperature. Constant growth conditions can be established by using a medium rich in nutrients, so that nutrient consumption does not diminish the

---

Quasispecies and RNA Virus Evolution: Principles and Consequences, by Esteban Domingo, Christof K. Biebricher, Manfred Eigen and John J. Holland. ©2001 Eurekah.com.

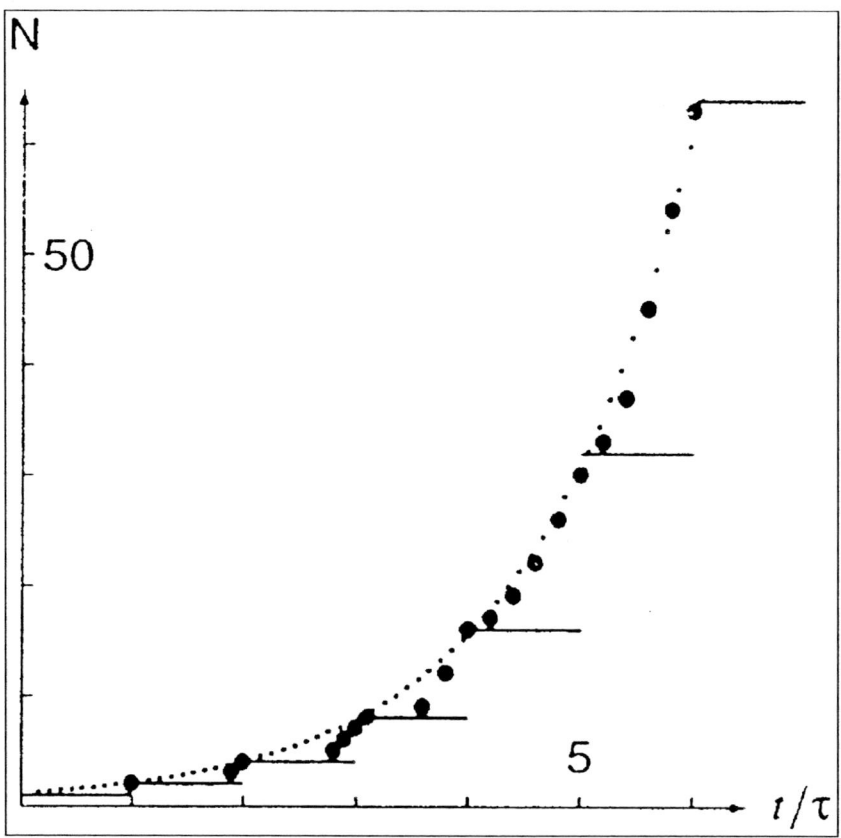

Fig. 4.1. Bacterial growth in solution. The points describe coherent growth; the lines show values for a fully synchronized growth; and the filled circles represent observed bacterial counts.

medium composition; vigorous stirring ensures that all bacteria have the same oxygen tension. The concentration cB of the bacterial population can be conveniently determined by measuring the turbidity of the culture medium. Accurate τ values can be obtained by using a mathematical trick: The logarithm of the bacterial concentration plotted against time gives at first a straight line, followed later by levelling off to a maximum value. The τ value is the inverse slope of the initial linear increase increase $\Delta \log(cB)/\Delta t$ divided by e.

Dynamical processes are described by ordinary differential equations. The rate of concentration change of a growing population is directly proportional to the number of parents, in our case proportional to the bacterial concentration. In mathematical symbols this is

$$dN/dt = \kappa N \text{ or } dcB/dt = \kappa cB, \quad (4.2)$$

where the proportionality constant κ is the specific growth rate. As long as κ is a constant, which is true for low bacterial concentrations, this differential equation corresponds to the formulas

$$N(t) = N(0)e^{\kappa t} \text{ or } c(t) = c(0)e^{\kappa t} \quad (4.3)$$

By comparing equations 4.1 and 4.3 we obtain $\tau = \ln 2/\kappa$.

We have discussed this growth law in detail because it shows the value of a theory: Quantitative relationships are established by an equation derived by plausible argumentation from observations. However, the observed growth curve shows the limitation of the theory: The law is only valid when special conditions are provided by the experimental setup. At higher bacterial concentrations the deviation from linearity becomes more and more noticeable as κ becomes dependent on the bacterial concentration; as the bacterial concentration increases, the oxygen tension of the solution and the growth rate decrease (Moser, 1957).

An empirical equation which fits the whole curve shown in Figure 4.2 can be derived by adding a negative term describing the levelling off at the maximum bacterial concentration c_{max}.

$$\mathrm{d}\,cB/\mathrm{d}t = \kappa(1 - cB/c_{max})\,cB \qquad (4.4)$$

When the bacterial concentration is small, the deviation from the exponential growth law is negligible, while the growth rate goes to zero when the concentration reaches its maximum. Equation (4.4) is empirical, not derived from theoretical argumentation and can not be used to draw theoretical conclusions.

It is possible to prolong the exponential growth phase by using a continuous flow reactor, feeding in fresh nutrients at the constant rate ϕ (Novick and Szilard, 1950a; Monod, 1950). The volume of medium V is kept constant by compensating the nutrient inflow by an equal outflow from the culture. Equation 4.3 becomes

$$\mathrm{d}\,cB/cB\,\mathrm{d}t = (k - \Phi) \qquad (4.5)$$

The exponential growth rate is slowed down by the flow; when the flow loss is higher than the growth rate, the difference in equation (4.5) is negative, resulting in the wash-out of the bacteria. What happens when flow and growth rates are exactly equal? The equation predicts a growth rate of zero and thus a constant bacterial concentration. If we try to verify this prediction experimentally, however, we obtain a puzzling result: The bacterial concentration makes irreproducible fluctuations in time, and after sufficient time spans the population dies out by an accidental washout.

How do we explain this result? We must take into account that in finite populations statistical fluctuations of the κ values are inevitable, leading to concomitant concentration fluctuations. Certain mutations may also affect κ values (see 4.8 below). Since κ is concentration-independent, fluctuations are not compensated.

For this reason, fermentors with constant nutrient feed rates are operated at concentrations where the growth rate is slightly concentration dependent. The combination of equations 4.4 and 4.5 leads to:

$$\mathrm{d}\,cB/\mathrm{d}t = (\kappa - \Phi)\,(1 - cB/c_{max})\,cB \qquad (4.6)$$

Now the feedback provided by the quadratic term holds the bacterial concentration constant, because fluctuations reducing the concentration speed up the growth while fluctuations increasing the concentration retard it.

Equation 4.3 can be applied to all sorts of organisms, for example lytic bacteriophages. The number of infective units in the population are determined by a plaque assay assuming that host cells are always in large excess. After an average eclipse time τ, an average burst of b mature viruses is released that infect new host cells. The specific growth constant is then $\kappa = \ln b/\tau$. What was noted above about synchrony also applies to this population. Because the phages multiply much faster than the hosts, some form of a flow reactor is required to keep the phage concentration at the exponential growth level. One apparatus used feeds a bacterial host solution into a reactor at a constant rate (Husimi et al, 1982; Biebricher et al, 1987). On the other hand, application of the equation 4.3 is not possible for eukaryotic viruses that infect multicellular

Table 4.1. Definitions of the parameters used to describe bacterial growth

N_i	Number of individuals of type i in the population	
c_B	Bacterial concentration in a fermentor	[L⁻¹]
τ_i	Average time between two bacterial divisions of type i	[s]
κ_i	Experimentally measured growth rate of type i in the exponential growth phase	[s⁻¹]
$\bar{\kappa}$	Growth rate of the total population	[s⁻¹]
i	Fraction of type i in the total population (type frequency);	
$\bar{x}_i$	Mutant frequency of type i in the quasispecies distribution	
Q_{ij}	Probability of producing type i per division of bacterial type j	
Q_{ii}	Probability of producing a correct copy of type i	
F_i	Relative fitness of type i, given by $F_i = \kappa_i - \bar{\kappa}$	

tissues; the complicated diffusion rates must be taken into account, and generally applicable growth equations are not yet available (parameters used in this section are defined in Table 4.1).

Selection of the Fittest

Equation 4.3 also applies when there are different types of bacteria in the population. Individuals that are indistinguishable by the experimental method in use are said to belong to the same "type". Depending on the experimental conditions, the types could be different species, or subspecies, or mutants. Again an experimental setup to keep the growth conditions constant must be used and no interactions between individuals of the population should take place. Furthermore, conversions from one type to another must be excluded. At high nutrient concentrations and low population densities, interactions become negligible and different types will grow independently of one another, i.e., each type grows as if others would not be present. The growth constant values κ_i are then characteristic for type i. The composition of the population will change with time if the κ_i differ from one another. For describing the population we introduce *relative concentration* variables $x_i(t) = c_i(t) / \Sigma_k^n {}_{=1} c_k(t)$. One obtains from Equation 4.3

$$dx_i(t) = \{\kappa_i(t) - \kappa(t)\}x_i(t) \quad (4.7)$$

where $\bar{\kappa}(t) = \Sigma_k \kappa_k x_k(t)$ represents the weighted average of all growth rate coefficients at time t.

The inherently nonlinear form of equation (4.7) (because changes with time) describes a selection process. Types with specific growth rates κ_i smaller than the average $\bar{\kappa}$ have negative growth rates and will be depleted in the population, while the population becomes enriched in types having rate coefficients larger than $\bar{\kappa}$. Concomitant with this change of the population composition is an increase of the average growth rate, and more and more types fall below the average, until eventually the population reaches a maximum growth rate with only type $\underline{m}$, i.e., the type with the maximal growth rate, surviving. Hence we have an extremum principle conforming to the expressions

$$\kappa(t) \to \kappa_{max} : x_m \to 1 : x_{i \neq m} \to 0 \quad (4.8)$$

The self-ordering process described by equation (4.8) is natural selection, and an immediate consequence of autocatalytic reproduction. Equation (4.8) is a quantitative description of natural selection and survival of the fittest, correlating the fitness of a type with its reproduction rate and its survival ($xm \to 1$) with the maximum reproduction rate.

Quantitative Molecular Evolution

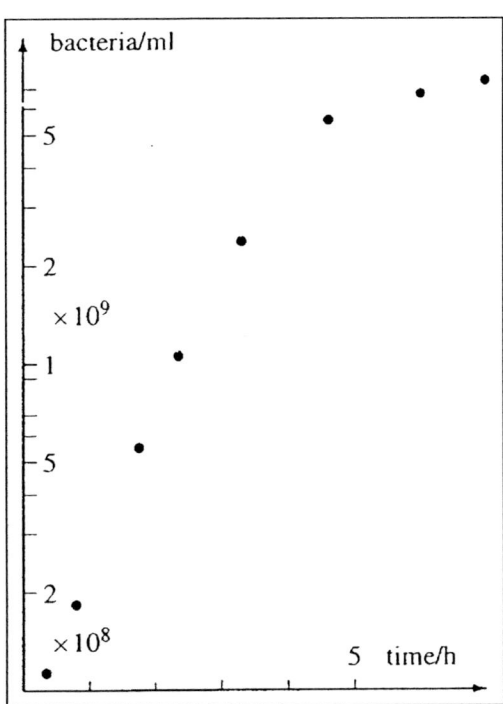

Fig. 4.2. Bacterial growth profile for *Escherichia coli* grown in rich medium under good aeration.

The validity of equation (4.8) can be tested by fermentor experiments. There are experimental difficulties: The times leading to final selection are long and one must introduce nutrient flux in order to ensure that the culture remains in the exponential growth phase. A constant flux is impossible, because the growth rate of the total population is changing. Monod (1950), and Novick & Szilard (1950a) independently invented such an apparatus, the turbidostat, to keep the total bacterial concentration constant.

While experimental evidence supports equation (4.8), it is often not adequate to describe natural selection: Darwinian behavior involves two forces, one that reduces the number of types in the population, selection, and another that increases the number of types, mutation.

Mutant Spectra and Sequence Space

Phenotypic Mutations

For a long time the only way to observe mutation was as emergence of new phenotypes. Nevertheless, Luria and Delbrück (1943) succeeded to show, in a quantitative experiment, that Darwin was correct in postulating an undirected mutational force. They measured the statistics of the emergence of bacterial mutants resistant to the bacteriophage T1 (phenotype *Ton*). There were two types: The sensitive (wild type) s and the resistant mutant r. Luria and Delbrück recognized that it is not correct to restrict the appearance of mutants to mutation alone: Mutants also reproduce and one must always take into account the differing selection among different mutants. If the "wild type" is far in excess, back mutation is unimportant and it is only necessary to consider the competition between wild-type and mutant. Under normal growth conditions the *Ton* phenotype is neutral, i.e., the specific growth rates of both mutant and wild-type are equal. According to equation (4.8), the change in the population composition by selection can therefore be neglected. Luria and Delbrück could show that the number of *Ton* progeny of a wild type bacterium scattered widely, depending on when the early mutation occurred, because the mutation is propagated to its progeny (Fig. 4.3).

Let us quantify and generalize the results of Luria and Delbrück. Mutations are not necessarily neutral: Most are detrimental, or even lethal, and negative selection rapidly limits their numbers. In a few cases, the mutation is advantageous and, once formed, the mutant is selected. We thus have to introduce into equation (4.8) an additional term reflecting mutation. Equation (4.8) associates selection with a deterministic parameter, fitness, while mutations inherently appear stochastically, regardless of their fitness. The wild-type s will produce mutant r (e.g., phenotype *Ton*) with a certain probability during its reproduction. When considering large populations and high mutant production probabilities, we may use a deterministic specific mutation rate Q_{rs} and obtain:

$$dxr / dt = \{Q_{rr} \bar{\kappa}_r - \bar{\kappa}\}x_r + Q_{rs} x_s \qquad (4.9)$$

When considering the growth parameter of type r, we must take into account that type r only makes correct progeny with the quality parameter Q_{rr}. In the Luria and Delbrück case we only have two types, and so $Q_{rr} = 1 - Q_{rs}$.

As long as the xr-values are small the mutation term prevails and one can (Figure 4.4) determine the mutation rate Q_{rs} from wild type to the *Ton* phenotype and also—by starting with a *Ton* phenotype—the reversion rate Q_{rs} (Novick & Szilard, 1950b). The linear increase of *Ton*-phenotypes showed that—under the conditions used—the mutation was neutral, i.e., the κ values of mutant and wild type were equal, and that the forward mutation occurred at a frequency about three orders of magnitude higher than reversion to the wild type. It was deduced, correctly, that the altered phenotype was caused by loss of a function; i.e., many mutations lead to the *Ton* phenotype while only one or at most a few mutations result in the reversion to wild type. Likewise, mutants with the *Ton* phenotype may themselves acquire additional disadvantageous or adavantageous mutations.

In chemostat cultures the emergence of phenotypic mutations could be followed for several weeks (Novick and Szilard, 1950b; Dykhuizen and Hartl, 1983). Erratic patterns appeared, with periods in which relative mutant concentration xr was increasing linearly and other periods where the relative mutant concentration decreased. The linear increase of the mutant concentration can be clearly interpreted with equation (4.9). In an almost homogeneous wild type population, mutants accumulate at a certain rate and as long the relative mutant population remains small, reversion can be neglected. But how can one explain the periods of decrease? This result is incompatible with our analysis, again because the model is too simplistic: There are not just two types in the solution. During the continuous growth of bacterial population, a large collection of mutants that do not show the *Ton* phenotype were formed, most of them disadvantageous, some of them neutral, but occasionally one that was advantageous. The advantageous mutants will be enriched in the population and eventually replace the old wild type, which goes extinct together with its mutation progeny. A set of mutants, some with the *Ton* phenotype is then rebuilt from the new dominant type. Likewise, mutants with the *Ton* phenotype may themselves acquire additional disadvantageous or advantageous mutations.

It may seem paradoxical that advantageous mutations can be found for a bacterium that had billions of years to evolve. Bacteria are adapted to a natural habitat that undergoes environmental changes all the time. The situation in a fermentor is quite different: There are no other competitors, and the medium is unnaturally rich. Biosynthesis of many important metabolites is thus dispensible and a loss of the genes required for synthesis saves the metabolic costs of intracellular production. Indeed, the *Escherichia coli* strains used in the laboratory are so 'domesticated' that they cannot survive in their natural habitat, the gut of warm-blooded animals.

Comparison of the properties of selected bacteria showed that their selective advantages were very small: The shape of their growth curves did not change noticeably. Equation (4.9)

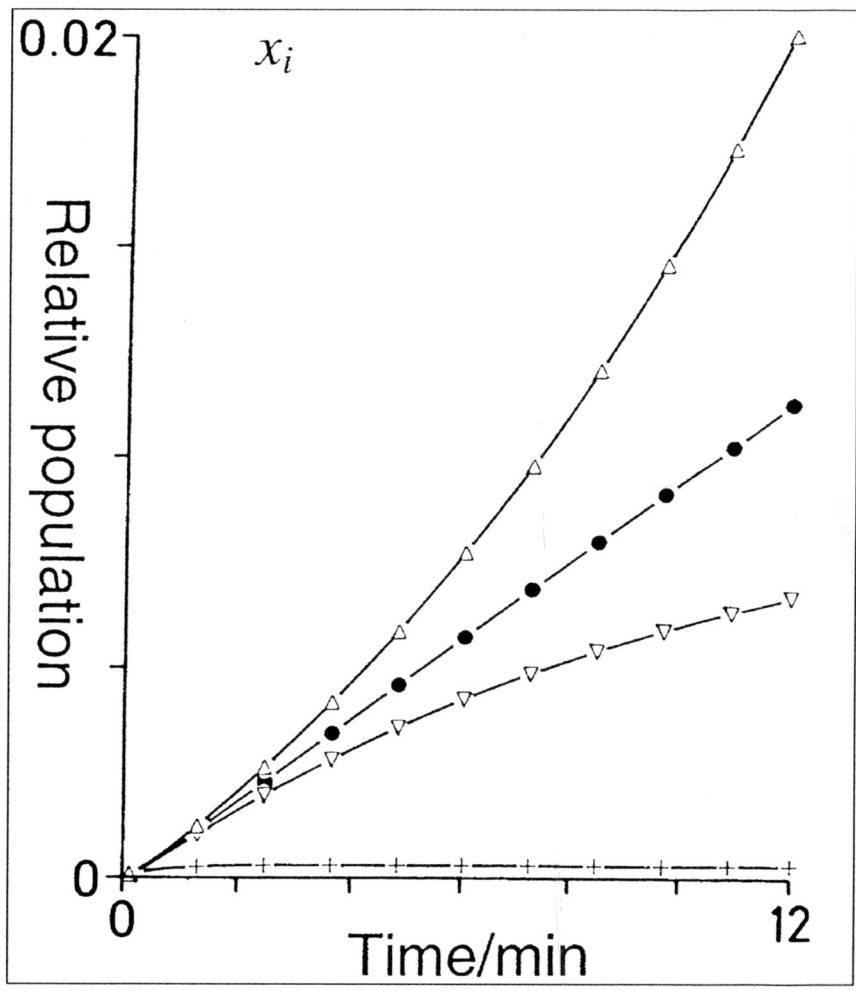

Fig. 4.4. The build-up of mutants in a population with time. The relative mutant frequencies of neutral mutants (•), advantageous mutants (Δ), disadvantageous mutants (∇) and lethal mutants (+) are plotted against time. Note that at very small mutant frequencies (i.e., in a homogeneous population of the master), the mutation rate can be determined from the slope. Only for neutral or nearly neutral the mutation frequency can be easily measured.

How should we then avoid having a population that already contains the mutants to be measured? A genetic trick solves the problem: Error propagation is avoided when we start with a type having a selection value of zero; then mutation can be measured in the absence of selection. Selection values of zero mean lethal mutations: How should we then prepare the population or analyze the phenotype? The solution is to use conditionally lethal mutants: While the mutants indeed do not grow under the conditions of the mutation experiment, they do so nearly normally under different conditions.

Benzer (1959, 1961) employed this technique to locate the mutant loci within a gene, obtaining a detailed mutation topography. For his measurements he used the so-called rII mutants which do not form infection plaques with the host *E. coli* strain K12 (containing the

Quantitative Molecular Evolution

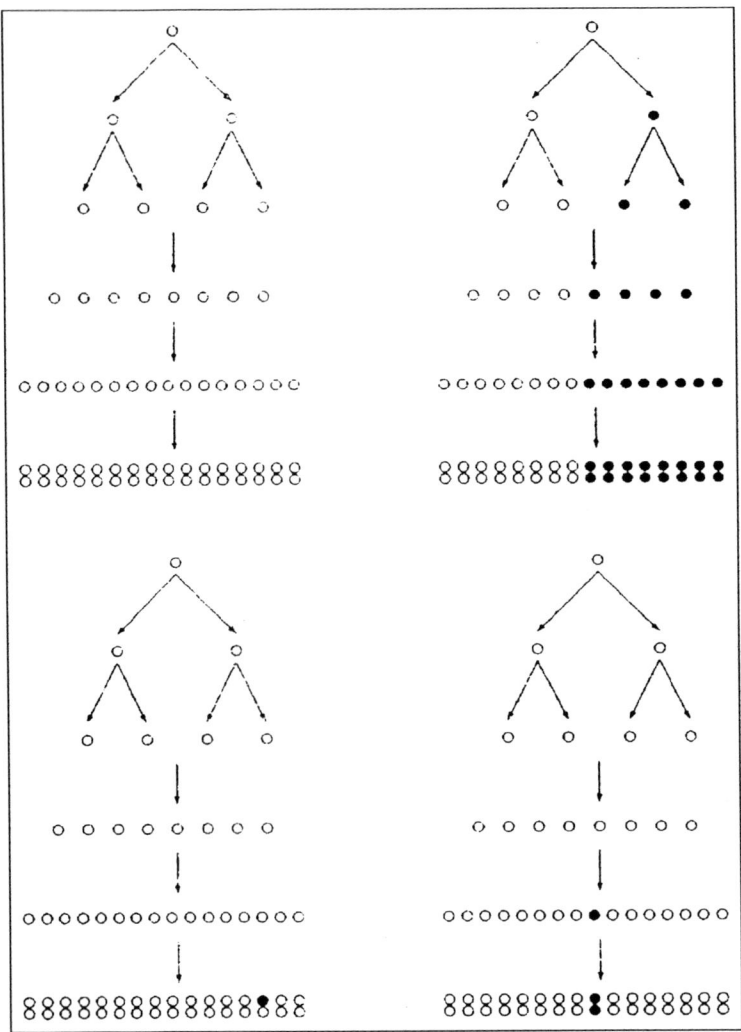

Fig. 4.3. Mutation experiment of Luria and Delbrück. Several samples of single bacteria were grown in nutrient broth, in absence of selection pressure, and the proportion of the *Ton* phenotypes (filled symbols) in the different populations were determined. The relative amount of *Ton* phenotypes scattered strongly, depending on when mutations occurred. In the scheme, the mutation frequency is greatly exaggerated to illustrate the principle.

tells us that tiny differences in the growth rate will suffice if the time spans are long and the populations high. In small populations, however, the growth rates scatter, and small differences in growth rate vanish in the scatter; i.e., the population composition seems to drift statistically. This neutral drift (Kimura, 1983) is thus a concept important for limited populations and variable environments, but not applicable if time spans and population sizes approach infinity.

Figure 4.4 shows that it is possible to calculate mutation rates, but the measurements are difficult: small differences in selection values make significant deviations from linearity unless the x_i values are very small. On the other hand, we need a large population to start the experiment:

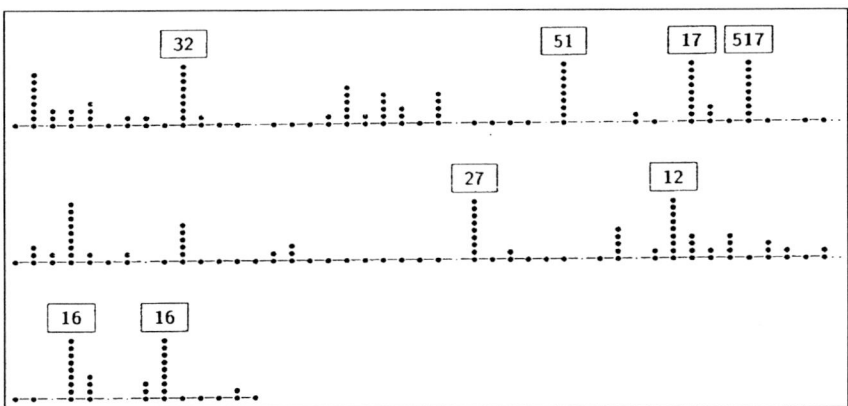

Fig. 4.5. The mutant topography of cistron b of the rII gene of the bacteriophage T4. Spontaneous mutations were crossed with other mutants and mapped according to the recombination probabilities. Mutants that could not compliment one another were considered identical. The locus of each mutation found is shown; the number of • indicate the mutant frequency found. At mutants that occurred at high frequencies, the mutant frequency number found is listed (disregarding the • symbols). Loci that originated from chemical mutagenesis not found in the spontaneous mutant spectrum are indicated by ·.

prophage λ), but give normal plaques with *E. coli* B cells. A large variety of such rII mutations could be prepared and their distance on the gene map determined. In pairwise double infection experiments with two different mutants, wild type progeny results from recombination between the mutants. Ordering the different pairs according to their recombination rate to wild type gave a rather precise map of the gene, which could be further refined by assuming that the recombination rates are inversely proportional to the distance of the mutation loci. Mutants not able to produce the wild type by recombination must be identical, and Benzer thus also obtained mutant frequencies, i.e., the number of mutants in the population. Benzer noted that the mutant frequencies differed widely for different loci on the rII gene, and he called the loci with particularly high mutant frequencies "hot spots" (Figure 4.5). This name seems to imply that the mutation frequency is enhanced at these points, and indeed that is the most plausible explanation, but a careful interpretion of the result reveals that other factors may also contribute to the mutant frequency. Benzer recognized some of the possible error sources and explicitly excluded "leaky" mutants, i.e., mutants that have a reduced but still measurable growth rate under nonpermissive conditions. On the other hand, unequal rates of recombination would not alter the mutant frequencies, but would cause merely distortions in the distance scale of the gene.

Sequence Heterogeneity

The prevalent feeling among microbiologists has been that species that grow vegetatively are genetically quite uniform, while sexually reproducing Mendelian species have high genetic diversity, their "gene pool". In 1977, this assumption was shattered by a genotypic investigation of virus populations (Domingo et al, 1978) and, independently, by a theoretical analysis (Eigen and Schuster, 1977). The shortcomings of phenotypic analysis are apparent: Most mutations do not change any phenotypic property. Mutations occur at the genotypic level, however, and all of them become apparent only upon examination of the nucleotide sequences of individual members of the populations.

The first genome sequenced was that of levivirus MS2 (Fiers et al, 1976). The nucleotide sequence was clearly defined, and when the RNA genome was digested with base-specific nucleases, a highly reproducible electrophoretic "fingerprint" pattern that could be used to identify the RNA unambiguously was obtained. As the sequence analysis of another levivirus, Qβ, progressed in the laboratory of Charles Weissmann (Billeter et al, 1969), the researchers became aware that the sequence might "drift" during growth passages of the virus. It was decided to reclone the phage for further sequencing studies. The result was shocking: The fingerprint pattern showed several alterations. Was it just bad luck of isolating a mutant by chance? Careful analysis of many clones showed that it was by no means a rare accident: Most clones had a few sequence deviations from the wild type, but the deviations appeared at different nucleotide positions. Multiple amplification passages of the virus, however, resulted in reappearance of the "wild-type" fingerprint. The Qβ population is thus highly heterogeneous, and the wild type is – even though by far the most frequent genotype – a minor fraction of the population. Other mutants average each other out, resulting in a defined wild-type sequence.

Eigen and Schuster (1977) came to the same conclusion by theoretical considerations. They started with an equation similar to equation (4.9), generalized for a large number of different mutants:

$$dx_i / dt = (W_{ii} - \overline{E}(t))x_i(t) + \sum_{k \neq i} W_{ik} x_k(t) \quad (4.10)$$

where W_{dm}, the mutation frequency, is the rate by which a daughter of type d is produced by the mother m, and $\overline{E}$ is the relative net increase of the total population. xi is the frequency (proportion) of mutant i in the population. The first sum term of equation (4.10) is the selection rate value; the second is the mutational backflow gain, i.e., the production of species i by mutational reproduction from all other genotypes.

Eigen and Schuster argued that for large population sizes, and after long growth times, a steady state is formed where each mutant (including the wild-type) occupies a constant part $\bar{x}_i$ of the population, i.e, $d \bar{x}_i dt = 0$. They called this complex population—which is equivalent to the "equilibrium population" obtained after several growth passages in the experimental work (Domingo et al, 1978)—a "quasispecies" and proposed that for the estimated low fidelities of the Qβ RNA replicase the "master" sequence, which has the maximum selection value and is in most cases the most frequent type, was present only as a small fraction of the quasispecies population. This prediction agreed with experimental findings with Qβ (Domingo et al, 1978), and it has been amply confirmed with many RNA replicons (Chapters 6 and 7). They showed also that the defined wild-type (consensus) sequence is an average of the total population, which coincides in most cases with the sequence of the master, which has the maximum selection value. Further analysis predicted that the chain length of the viral RNA contains the maximum information that can be stably maintained with the fidelity levels displayed by the replicase. This "error threshold" plays an important role for the genetic organization of viruses and for new antiviral strategies, and will be revisited in more detail later (later in this Chapter and in Chapter 8).

The Sequence Space

The large sequence heterogeneity of RNA viruses has been seen in all RNA viruses investigated (see Chapter 6). Normally, "mutant spectra" are described by aligning the mutant sequences with that of the "wild type" (or zero mutation class). The spectrum contains many one-error mutants, but also others which have more than one base exchanges, expressed as a larger Hamming distance d_{iw}. Normally, a Hamming distance (Hamming, 1980) is defined for genetic analysis as the number of nucleotide differences between two sequences; it would be appropriate to refine it by a weighting function that takes into account differences in mutation probability, e.g., to distinguish base transversions from base transitions.

Some mutants may have high Hamming distances from the wild type. It is unlikely that such mutants have been formed directly from the wild-type in a single replication cycle. Instead, such sequences contain information about the kinship of the mutants. A topography to illustrate this kinship is the sequence space (Eigen, 1992). How can such a space be represented? A plane suffices to represent sequences of length two, a cube for representing sequences of length three. One-error mutants are then connected by an edge, two-error mutants by diagonals across planes, and the three-error mutant at the opposite corner of the cube. Continuing beyond chain length three requires a v-dimensional hypercube, which surpasses our steric imagination. However, three essential features of the high-dimensional point space are apparent. (For simplicity, we consider only a sequence space of binary symbols; the space can be expanded to accomodate the four symbols of the genetic alphabet):

- The storage capacity, i.e., the number of corners of the hypercube, increases rapidly with increasing dimension, i.e., as $2v$. A 360-dimensional space (representing a nucleic acid sequence with 180 positions) would suffice to map the whole universe (viewed as a sphere with a diameter of about 10 billion light years) with a resolution in the $Å^3$ range, assuming as volume of the universe 10^{108} $Å^3$. Realistic genome lengths imply hyperastronomically large numbers.
- Despite the large capacity, the shortest path to any point in the space is short indeed: v steps suffice to reach the farthest point. If each step is unguided, however, the chance of reaching a distant target by a random walk is practically nil.
- The large connectivity among all points of the v-dimensional hypercube is intuitively obvious. In a v-dimensional space every point has v nearest neighbors, and $\binom{v}{d}$ neighbors within a distance d. In the above-described space, every member has 10^9 neighbors within a distance of 5 paces.

We now use the concept of sequence space for deepening our understanding of the quasispecies.

The Quasispecies

No dimension in the sequence space is "favored", so we could have used our above arguments just as well for a v-dimensional plane. A vertical coordinate can be used to represent the selection value, which provides a force directing a walk in sequence space. In this way a "fitness landscape" is obtained. While gravity would guide the walk towards lower heights, the selection force drives the walk uphill. If random heights were assigned to the different points on the sequence plane, a natural landscape would not be obtained. The distribution of heights in a natural landscape is not random; the altitudes of near-neighbor points usually are not very different.

However, while models are useful for testing possible properties of a phenomenon, it is indispensable to show that the inferred property is indeed observed. This applies to the fitness landscape. It is certainly much more rugged than a natural landscape, because it is not continuous. By far the largest parts of the fitness landscape are lowlands with height zero; it is extremely improbable to hit an elevated point by chance. On the other hand, natural populations contain many few-error mutants that are viable or even (almost) neutral, i.e., they occupy elevated points in the fitness landscape, the highest peak of which corresponds to the master sequence for a defined environment. By directed mutation experiments, it can be shown that the further we protrude from the master into the surrounding sequence space, the more difficult it becomes to hit a success. Experimentally, the cross-section through a fitness mountain is by no means anything as simple as a Gaussian function. A typical tour in the fitness landscape from the lowlands to the master peak would thus not be a monotonous ascent, but would require ascents and descents.

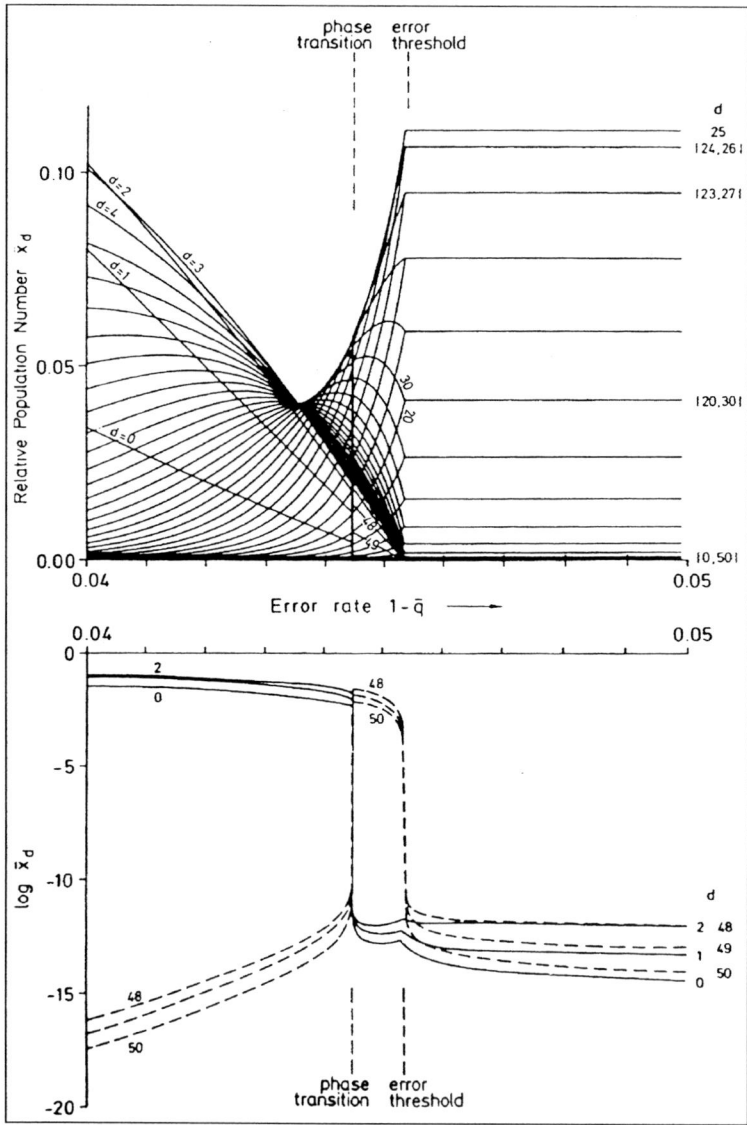

Fig. 4.6. The target of natural selection. The phase transition-like character of natural selection is demonstrated in this computer simulation (data from Swetina J, Schuster P. Self-replication with error—a model for polynucleotide replication, Biophys Chem 1982; 16:329-345). Assumed is a binary sequence of complexity $\upsilon = 50$. Mutants in the same error class are assumed to have degenerate selective values. The master {0} has the highest selective value W_0, the (single) 50-error mutant {50} has $0.9 \times W_0$, the 50 different 49-error-mutants have $0.5 \times W_0$, all other mutants have $0.1 \times W_0$. In the simulation, the relative population numbers of the different error classes x_d (obtained by summing over all x_i-values of their N_d members) are plotted against the error rate. In the lower plot a semilogarithmic plot is shown to show small contributions. Two phase transitions are seen. At low mutation rates $1-\bar{q}$ the sequence {0} is the winner of the competition; the relative concentration of the near-neutral {50} competitor is very small. With increasing error rate, all mutants increase at the cost of {0}. At a certain error rate, there is a sharp 'phase transition' where the slightly inferior sequence {50} takes over by virtue of its better mutant environment and x_0 drops to very low value.

It is interesting to inquire how a quasispecies is populated. To do so, another landscape on the sequence plane must be constructed using absolute or relative population values as heights. In contrast to fitness, which is an intensive quantity, population is an extensive quantity. Population space can be studied experimentally far more easily than any fitness landscape; indeed, the previously described sequence heterogeneity measurements (Domingo et al, 1978) as well as Benzer's experiments derived all their conclusions from investigating population landscapes. Full analysis of a real population landscape would require the sequence analysis of a statistically significant number of mutants, a task still not feasible with normal resources. However, the experiments described in the next Chapter and many other data allow us to draw some conclusions.

- We see clearly that the population landscape and the fitness landscapes are not congruent! Therefore, Darwin's theory is not a trivial tautology boiling down to survival of the survivors. A second trivial assumption is also clearly refuted: The population landscape is poorly correlated to mutation rates, and it is not possible to derive mutation rates from mutant frequencies. Equation (4.10) clearly tells us that we have three correlated parameters: relative population (mutant frequency), fitness (selection rate value) and mutational gain (mutation rates) and two of them must be known to calculate the third.
- Comparison of the fitness landscape and the population landscape reveals that the fitness landscape is much more "rugged": a single step may lead precipitously to zero height, while a second, compensating step may lead to a comparable fitness. On the one hand, the population landscape is less rugged because the mutation term smoothes it; there is no nonpopulated locus in the immediate neighborhood of a highly populated one. On the other hand, relatively small positional differences in the fitness landscape can cause dramatic effects in the population landscape.
- The population space does not monotonously decrease progressing away from the master sequence. As in a natural landscape on earth, there are hills with smaller elevations grouped around the highest peak. These hills are the equivalent of a clan in a normal population, because they comprise individuals of close kinship. The quasispecies is thus divided into subpopulations that compete normally one with another, and the mutation terms in equation (4.10) are dominated by contributions from members of the same clan.
- The highest mountain peak in the fitness landscape is the master. It is not necessarily the highest peak in the population landscape because of the contributions of the nearest neighbor. The wild type consensus or average sequence is the gravity center of the mountain in the population landscape which also does not necessarily share the same locus in sequence plane with the master or the most populated mutant.

How should we visualize a quasispecies distribution (Domingo et al, 1995)? One way is to imagine it as a continuous cloud in sequence space, denser in the interior, thin and fuzzy at the rim of the cloud. The edges are irregular: Some parts protrude far out, while in some directions the void begins after a few steps. This cloud in sequence space may not be the only one; many different species of similar genome length may thrive in the same regions. The clouds are not

---

The logrithmic plot (lower part) emphasizes the sharp transition and the strong selection. Further increases of the error rate lead to a drop of x_{50} in favor of its mutant spectrum, until eventually a second (and final) error threshold is surpassed where the selection forces do not suffice anymore to stabilize any master. All members of the sequence space are then equally represented ($x_i = 2^{-50}$). Since N_{25} (the number of individual sequences with a Hamming distance 25) is highest of all N_d (Table 4.2), x_{25} dominates. Selection hence may be viewed as a kind of condensation of information in sequence space: Exceeding the error threshold means that the information volatilizes through an error catastrophe, the error threshold being an analogue of an evaporation temperature. (Courtesy of Drs. Shcuster and Setina).

Table 4.2. Definition of parameters used in this Chapter

v_i	The genome length of type i or the sequence complexity.
d_{ij}	The Hamming distance between two sequences i and j, i.e., the number of positions at which the genomes differ.
μ	The number of different symbols in the sequence (4 for nucleic acids).
q	The fidelity, i.e., the probability of inserting the correct nucleotide.
$1-q$	The error rate, the probability of misincorporation.
Q_d	The relative frequency of misincorporation into a sequence of length v.
N_d	The number of different copies in the d-error class.
Q_{ij}	The probability of producing type i by reproduction of type j.
$Q_{ij} \equiv Q_d/N_d = q^v \{(q^{-1}-1)/(\mu-1)\}^{d_{ij}}$	

uniformly distributed in sequence space as a phylogenetic kinship of many species; instead, the clouds are separated by large void distances, and the sequence barrier thwarts any possibility of transfer from one cloud to another. There are large numbers of mountainous regions in the fitness landscape, where the population landscape has no equivalent: some species are extinct and others not created yet.

Exploration of sequence space is more efficient for RNA viruses than for cellular organisms. Let us illustrate it by an example: The phenotypes of different species of the *Leviviridae* family or of the rhinoviruses are practically indistinguishable one from another: Host range, gene organization, plaque morphology, reproduction rates, electron-microscopical appearance and the properties of the proteins are very similar. It is thus likely that they evolved from one ancestor. Nonetheless, their genotypes are so different that even an alignment of genomic sequences from different representatives is often difficult. This is in sharp contrast to cellular organisms, where the phenotypes differ enormously while the genotypes show considerable sequence conservation. The fitness landscape of *Leviviridae* is much more extended than the population landscape: their phylogenetic inheritance demands that there is a wide mountainous region in the fitness landscape connecting all *Leviviridae* genotypes, while the actual population of a virus clone occupies but a tiny fraction of them. In the laboratory the wild types of phage isolates can be amplified for many generations without much genetic drift, indicating a defined fitness peak under normal laboratory conditions (cf. "punctuated equilibrium" Gold and Eldredge, 1977). Environmental conditions vary more in the natural habitats, and the sequence drift observed in geographically different isolates of one virus reflects these altered conditions (Chapter 6).

Speculations are useful, but to make a speculation scientifically interesting, it must be free from inherent contradictions and in agreement with observations. The more speculative a proposal is the scarcer are the possibilities to test it by experiments. However, theories do provide tools that can expose inconsistencies; one can use mathematical laws, add plausible assumptions to simplify them, and perform computer simulations. While neither computer simulations nor experimental evidence can deliver the ultimate proof of a theory, both help to exclude alternative explanations, raising the probability considerably that a proposed theory will stand the test of time. Regarding quasispecies and fitness variations, an increasing wealth of experimental evidence (discussed in Chapters 5-7) supports the theoretical concepts expressed mathematically in previous paragraphs. Table 4.2 summarizes the parameters used in this section.

The Error Threshold

In Chapter 3 we described the fidelity of template-instructed replication, and explained how it can be measured. In cellular organisms, with their very long genome chain lengths, in vivo fidelity is orders of magnitude higher than in vitro, because there is a large post-replication apparatus that finds and corrects errors or chemical damage in the DNA. "Hot spots" with high DNA mutation rates are thus often caused by a failure of the "repair" systems to find and correct errors at specific loci. RNA viruses are usually single-stranded and post-replicative repair is not possible. The mutation rates during replication in vitro and in vivo are roughly in agreement. The use of uniform mutation rate values allows the calculation of mutation spectra according to the Poissonian distribution:

The numbers calculated (Eigen and Biebricher, 1988) tell us that Q_{ij} (see Table 4.2 for the meaning of different symbols) drops sharply with increasing Hamming distance and increasing sequence length, while Q_d is nearly independent of the sequence length and moderately dependent on the Hamming distance. In a normal laboratory sample containing 10^{12} viruses, mutants can be found with Hamming distances of up to 15. However, to find just one specific sequence with 10 positions out of 3000 altered would require a population of some 10^{30} viruses, corresponding to 17 million metric tons of viral RNA.

The higher the error rate per nucleotide and the chain length, the smaller is the probability to get offspring that are identical to the parent and to find the master sequence in the population. Eventually, the dispersing force of mutation takes over and can not be compensated by the focussing force of selection. There is an "error threshold" where the genotype and its information content can no longer be stably maintained (Eigen, 1971).

$$\nu max < \ln\sigma_0/(1 - \bar{q}) \qquad (4.11)$$

In this relation, the maximum chain length of a genome that can be maintained, ν_{max}, is inversely proportional to the average error rate per nucleotide and replication round, $\bar{q}$. The factor σ_0 gives a measure for the average selective advantage of the master in relation to its mutant competitors. Obviously, this quantity must be larger than unity, because selection is needed to filter out correct copies. Computer simulations (Swetina and Schuster, 1982) show that the mutant spectrum gets broader with increasing error rate until an error catastrophe at a critical error rate causes information to evaporate. Measurements of the chain lengths and the replication error rates of RNA viruses show that the genome lengths of RNA viruses are close to the maximum that can be maintained at the error rates measured for RNA replication. The fact that the majority of RNA virus particles are not infectious even though physical defects of the nonviable viruses can not be detected suggests that viral populations operate near to the error threshold (see also Chapter 8). The inequality (4.11) also shows that parts of the genetic information that do not contribute to the selection will eventually degenerate.

Measurements of error rates have shown that while RNA viruses are unique in their requirement of very short genomes, they share with the majority of organisms the fact that they thrive right at their predicted error threshold. Only bacteria and some DNA viruses seem to be exceptions; their measured replication fidelity apparently is higher than expected from their genomic complexity. However, bacteria often have to survive harsh conditions with high chemical mutagenesis. It is known that under stress, e.g., when the genome is damaged, the fidelity of bacterial replication drops by several orders of magnitude. Some DNA viruses usurp for their own replication the host replication apparatus, which is adapted to a much higher genome length. Other, more complex DNA viruses encode their own DNA polymerases, usually also including a proofreading activity (Chapter 7). It is not clear yet whether some complex DNA viruses develop additional devices to increase their genetic diversification rate, e.g., by coding for nonspecific recombination enzymes, or by interfering with cellular repair functions.

The fitness of a type defines its evolutionary success, i.e., the trend in which its population size may develop. Equation (4.11) shows the many contributions to this quantity. Computer simulations show clearly that—at error rates near to the error threshold—it is not necessarily the individual with the highest selection value that is selected (Swetina and Schuster, 1982; Eigen, 1986). The selection values of close kinship also contribute, and a small clan with an isolated highly fit leader is usually at a disadvantage compared to a large clan with smaller but more broadly distributed selection value. If, in this simulation, the error rate is *ceteris paribus* continuously decreased, a second "phase transition" is observed at a defined error rate below which the small clan with the higher singular selection value is favored (Swetina and Schuster, 1982). A number of studies on fitness variations of RNA viruses both in cell culture and in vivo are extensively discussed in Chapter 7.

References

1. Atwood KC, Schneider LK, Ryan, FJ. Selective mechanisms in bacteria. Cold Spring Harbor Symp Quant Biol 1951; 16:345-55.
2. Benzer S. On the topology of the genetic fine structure. Proc Natl Acad Sci USA 1959; 45:1607-1620.
3. Benzer S. On the topography of the genetic fine structure. Proc Natl Acad Sci USA 1961; 47:403-415.
4. Biebricher CK, Eigen M, Gardiner WC et al. Modeling studies of RNA replication and viral infection. In: Complex Chemical Reaction Systems; Mathematical Modelling and Simulation. Proceedings of the second workshop, Heidelberg, Aug. 11-15 1986 (Warnatz J, Jäger W, eds.). Berlin: Springer-Verlag 1987:17-38.
5. Billeter MA, Dahlberg JE, Goodman HM et al. Sequence of the first 175 nucleotides from 5' terminus of Qβ RNA synthesized in vitro. Nature 1969; 224:1083-1086.
6. Domingo E, Holland JJ, Biebricher C et al. Quasispecies: The concept and the word. In: Gibbs A, Calisher C, García-Arenal F, eds. Molecular Basis of Virus Evolution. Cambridge University Press, 1995; 171-180.
7. Domingo E, Sabo DL, Taniguchi T et al. Nucleotide sequence heterogeneity of an RNA phage population. Cell 1978; 13:735-744.
8. Dykhuizen DE, Hartl DL. Selection in chemostats. Microbiol Rev 1983; 47:150-168.
9. Eigen M. Selforganization of matter and the evolution of biological macromolecules. Naturwissenschaften 1971; 58:465-523.
10. Eigen M. Self-organization of matter and the evolution of biological macromolecules. Naturwissenschaften 1971; 58:465-523.
11. Eigen M. Steps towards life. Oxford: Oxford University Press, 1992.
12. Eigen M, Schuster P. The hypercycle—a principle of natural self-organization. Part A: Emergence of the hypercycle; Naturwissenschaften 1977; 64:541-565.
13. Eigen M, Biebricher CK. Sequence space and quasispecies distribution; In: Domingo E, Ahlquist, Holland JJ, eds. RNA Genetics Vol. III: Variability of RNA genomes. Boca Raton: CRC Press 1988; 211-245.
14. Fiers W, Contreras R, Duerinck F et al. Complete nucleotide sequence of bacteriophage *MS2* RNA: primary and secondary structure of the replicase gene. Nature 1976; 260:500-507.
15. Gould SJ, Eldredge N. Punctuated equilibria: The tempo and mode of evolution reconsidered. Palaeobiology 1977; 3:115-151.
16. Hamming RW. Coding and information theory. Prentice-Hall Englewood Cliffs, N.J. 1980.
17. Husimi Y, Nishigaki K, Kinoshita Y. Rev Sci Inst 1982; 53: 517-523.
18. Kimura M. The neutral theory of molecular evolution. London: Cambridge University Press 1983.
19. Luria SE, Delbrück M. Mutations of bacteria from virus sensitivity to virus resistance. Genetics 1943; 28:486-491.
20. Monod J. La technique de culture continue. Théorie et applications; Ann. Inst. Pasteur Paris 1950; 79:390-410.
21. Moser H. Structure and dynamics of bacterial populations maintained in the chemostat; Cold Spring Harbor Symp Quant Biol 1957; 22:121-137.
22. Novick A, Szilard L. Description of the chemostat. Science 1950a; 112:714-716.
23. Novick A, Szilard L. Experiments with the chemostat on spontaneous mutations of bacteria. Proc Nat Acad Sci USA 1005b; 34:708-719.
24. Schuster P. The physical basis of molecular evolution. Chem. Scr. 1986; 26B:27-41.
25. Swetina J, Schuster P. Self-replication with error – a model for polynucleotide replication. Biophys Chem 1982; 16:329.

CHAPTER 5

Darwinian Evolution of RNA in Vitro
Extracellular Darwinian Experiments

In the previous Chapter we described how quantitative measurements have led to studies of Darwinian evolution in action. Simplifications and abstractions were required, and experimental systems in which reproducible and constant environmental conditions can be established are of great value. An ideal system should have only a few biochemical steps for translating a genotype into a phenotype and a short genome for keeping the sequence space small and for conveniently sequencing the entire genome, to define the genotype unambiguously. Finally, one needs a high mutation rate to observe evolution at laboratory scale time periods. Table 5.1 summarizes key parameters of a number of biological systems that have been employed to approach evolutionary problems, starting with the garden pea used by Mendel, progressing to the *Drosophila* system used in the first half of the XXth century, and onward up to bacteria, viruses and finally RNA molecules.

The first and still one of the best-suited experimental systems for studying evolution in vitro was devised by Sol Spiegelman (1971). Soon after discovering a procedure for replicating RNA in vitro (Spiegelman et al, 1965), he began to exploit its potential for the study of molecular evolution. In 1967 he and his coworkers (Mills et al, 1967) published a classical work entitled "An extracellular Darwinian experiment with a self-duplicating nucleic acid molecule". They wrote: "It is of great interest to design an experiment which attempts an answer to the following question: 'What will happen to the RNA molecules if the only demand made on them is the Biblical injunction, multiply, with the biological proviso that they do so as rapidly as possible?'". They started their experiment by providing a growth medium containing the monomer precursors and Qβ replicase as the enzyme to replicate infectious Qβ RNA. After a few minutes the reaction would stop because as RNA had accumulated the precursors were consumed; to circumvent this, they serially transferred aliquots of RNA into fresh growth medium. Under these conditions, because RNA was neither translated nor packed into virions it was plausible that much of the genome information was dispensable. Indeed, after just 5 transfers the synthesized RNA was already no longer infectious. The rate of RNA synthesis increased after a few transfers, and the incubation periods for the transfers could be shortened. After 75 transfers, the RNA was analyzed by the techniques then available. From the sedimentation rates it was estimated that the final replicating RNA had retained only about 17% of the chain length of viral Qβ RNA, yet it showed a 2.6-fold enhanced nucleotide incorporation rate. Therefore, the replication rate per RNA molecule was estimated to be about 15-fold compared to Qβ RNA. Its base composition differed significantly from that of viral RNA; its A:U and G:C ratios differing from unity indicated a single-stranded product, which was confirmed by its sensitivity

---

Quasispecies and RNA Virus Evolution: Principles and Consequences, by Esteban Domingo, Christof K. Biebricher, Manfred Eigen and John J. Holland. ©2001 Eurekah.com.

Table 5.1. Some biological systems used in evolutionary research

Species name	Genome size [bases]	Generation time [s] size	Typical population size	Mutation rate (subst./nucleotide)
Pisum sativum	1×10^{10}	3.15×10^7	5×10^2	$10^{-5} - 10^{-12}$
Drosophila melanogaster	2.6×10^8	1.2×10^6	5×10^3	$10^{-5} - 10^{-10}$
Escherichia coli	3.4×10^6	1.0×10^3	10^{11}	$10^{-4} - 10^{-8}$
Phage Qβ	4.2×10^3	240	10^{14}	3×10^{-4}
MNV11 RNA	86	40	10^{16}	3×10^{-4}

to ribonucleases. Spiegelman and colleagues concluded that a new RNA "variant" had gradually evolved from the Qβ RNA.

Experiments starting with Qβ RNA, or one of the variants, and applying different selection pressures "revealed an unexpected wealth of phenotypic differences which a replicating nucleic acid can exhibit", e.g., growth at low or strongly biased nucleotide precursor level or resistance against replication inhibitors (Levisohn and Spiegelman, 1969; Saffhill et al, 1970). New phenotypic properties were shown to be passed on to descendants grown in the absence of the selection pressure, which can only be explained in terms of mutation and selection. The selection conditions and the products are compiled in Table 5.2

Spiegelman's experiments were a milestone in the study of evolution, showing for the first time that molecules able to sustain self-reproduction show Darwinian evolution just as living organisms do. Previously, phenotypic alterations were supposed to be expressed only by altered proteins. The isolation of adapted RNA mutants under conditions where the replicase could not be altered shows that modification of an RNA interacting with a preexisting protein is an efficient alternative strategy of evolution (Orgel, 1979).

Unfortunately, the limited knowledge at that time of the RNA replication system and the lack of defined homogeneous replicase preparations prevented a clear-cut interpretation of what was happening at the molecular level. Indeed, many results were puzzling and difficult to reconcile with a gradual evolution (Biebricher, 1983):

- Different independent isolates of the variant RNAs had indistinguishable properties, e.g., the same molecular weight and the same base composition. Mutations, being stochastic and irreproducible events, should lead to different emerging variants.
- Sequence homologies or hybridization between Qβ RNA and the variant RNAs were not established. It could not be excluded that the variants orginated by another event or from another source, since replicase preparations at that time were partially fractionated, crude extracts from infected bacteria.

It is a pity that such a central experiment has not been repeated under more stringently defined conditions and with careful investigation of the evolution intermediates. Spiegelman and colleagues suggested it, remarking: "The experimental situation provides its own paleology; every sample is kept frozen and can be expanded at will to yield the components occurring at that particular evolutionary stage." (Mills et al, 1967).

Spiegelman's experiments provoked mixed reactions, ranging from enthusiasm to irritation. Some virologists considered the experiment "the search for the most beautiful carcass" and the resulting variants as "nonphysiological". This is of course true, because the resulting variant has lost infectivity, together with most other capabilities of the viral RNA. As with genetic experiments using bacteria, the observed evolution is essentially degenerative. However, evolution experiments in vitro abstract from the enormously complicated networks acting

Table 5.2. Properties of variants of Qβ RNA

Variant	Precursor	Number of transfer	F_{dil}	Conditions	Phenotype
V-1	Qβ	74	12	standard	fast growth
V-2	*Qβ	17	10^3-10^8	standard	could be cloned
V-3	V-2	?	?	16 µM CTP	grows at limiting [CTP]
V-4	Qβ	10	10^5-10^{11}	16 µM CTP	indistinguishable from V-3
V-6	V-4	40	10^4	5 µM CTP	as V-4
V-8	V-6	16	10^4	12 µM NTP	grows at limiting [NTP]
V-9	V-8	19	10^4	240 µM TuTP	resistant to tubercidin
V-40	V-2	108	11	40 µg/ml EtBr	resistant to EtBr

in vivo to focus attention on one basic step, RNA replication. The phenotype is thereby clearly defined as the ability of an RNA to direct the replicase to replicate it; there is no need for gene expression by translation and for regulated synthesis of precursors. Darwinian evolution is not necessarily associated with higher complexity of the product of evolution; it does not have an inherent teleology.

In order to carry out quantitative RNA evolution experiments in vitro the following conditions must be met:
1. The replicase and the substrates must be meticulously pure, in particular, free of contaminating templates.
2. A defined replicating RNA species, if possible, of short chain length, must serve as template.
3. During the experiments, evolution intermediates must be isolated and characterized.
4. The genotypes of the starting templates, the evolutionary intermediates and the selected product must be compared. An unequivocal sequential kinship of these RNAs must be shown.

Spiegelman and coworkers were able to isolate and characterize an RNA species that was suitable for evolution experiments under defined conditions (Mills et al, 1973). It was named "midivariant" or MDV-1, because they assumed that the sequence was derived from the Qβ genome itself. With a chain length of 221 nucleotides, it was larger than many other replicating RNA species but shorter than the variants listed in Table 5.2. Sequence analysis of MDV-1 RNA indeed revealed sequence relationships with Qβ viral RNA; however, since about as much or even higher homologies can be found to other RNA sequences, that homology is inconclusive.

A large variety of short-chained RNA species have been isolated—essentially by incubating replicase with substrates in the absence of extraneously added template—and sequenced (Schaffner et al, 1967; Biebricher and Luce, 1993). Their chain lengths range from 25-250 nucleotides. We shall discuss their origin later, and consider first the chemical kinetics of the replication process itself.

The Kinetics of RNA Replication

For quantitative studies of RNA evolution a mathematical description is indispensable. For more complex replicons such as viruses, discussed in following Chapters, kinetic parameters are difficult to define and to express in mathematical terms. The parameters used in the following sections are given in Table 5.3.

RNA Growth Rates

Quantitative replication rate studies have been done by measuring the rate of incorporation of nucleotides into RNA. A typical incorporation profile is shown in Figure 5.1. Three phases of replication can be distinguished:
1. An exponential growth phase, where enzyme is in excess relative to the template, and the RNA concentration is amplified exponentially.
2. A linear growth phase during which the replicase is saturated with template.
3. Eventually, for reasons similar to those governing bacterial growth curves, the RNA concentration levels off to a maximum value.

Kinetic studies of the replication process support the replication mechanism described in Chapter 3. The RNA template binds specifically at its 3' terminus to the replicase enzyme, and the replicase synthesizes a complementary antiparallel replica strand. After the replicase reaches the 5' terminus of the template, the replica is polyadenylated at its 3' terminus and released. The remaining inactive template-replicase complex dissociates slowly into its components; this step is usually the slowest step in the cycle. In the linear phase, recycling of the replicase is rate-limiting, while excess replicase is always available in the exponential phase (Biebricher et al, 1983).

Several consequences of these features have to be considered when interpreting the selection of self-replicating RNA species:

- RNA molecules do not replicate strictly autocatalytically; there are two complementary RNA strands that replicate one another cross-catalytically. A mutation in one strand results in a corresponding mutation in its complementary strand. Hence, the genotype comprises both complementary sequences.
- Because the template is released much later than the replica, the duplication kinetics are not as simple as with dividing bacteria. The growth characteristics are more closely analogous to Fibonacci's two-step growth system, illustrated by the increase in the number of rabbits reproducing and maturing over equal time intervals. At large population sizes, the series of Fibonacci numbers grows exponentially (Eigen et al, 1991). Because excess replicase is always available in the exponential phase, the overall replication rate in the exponential phase is higher than in the linear phase, where recycling of the replicase is rate-limiting.
- Single strands fold to complicated secondary and tertiary structures that strongly influence the rates of the initiation, elongation and termination steps. For evolution studies, this is highly advantageous, because all parts of the sequence are important for selection. If only parts of the sequence (e.g., initiation signal sequences) would determine the replication rates, selection degeneracy in the other sequence parts would result, and a stable mutant distribution could not be built.

The steps involved in RNA replication have been described in Chapter 3. We can choose conditions which allow the reaction dynamics to be described in simplified form. Mutations are disregarded, and saturating, buffered monomer concentrations and high enzyme concentrations can be assumed. Under those conditions, all steps required for replica synthesis may be combined into a single first-order elongation step. If all steps are considered to be irreversible, the cross-catalytic cycle depicted in Fig. 5.2 is obtained. If plus and minus strand cycles behave similarly, the mechanism reduces to a single effective autocatalytic cycle. Indeed, optimized fast-growing RNA species are observed to have about equal concentrations of plus and minus strands, suggesting similar rate constant values for the complementary cycles.

Table 5.3. Definition of the parameter used for RNA replications studies

	Concentrations	
$[^iI]$	Concentration of free single-stranded RNA of type i	[mol/l]*
$[^iEI]$	Concentration of active replication complexes with template of type i	[mol/l]
$[^iIE]$	Concentration of inactive replication complexes with template of type i	[mol/l]
$[E]$	Concentration of free enzyme	[mol/l]
$[^iE_c]$	Total concentration of template strands of type i complexed to enzyme; $[^iE_c] = [^iEI] + [^iIE]$	[mol/l]*
$[E_0]$	Total concentration of enzyme; $[E_0] = [E] + \Sigma_i[^iE_c]$; standard value 2×10^{-7}	[mol/l]**
$[^iI_0]$	Total concentration of template strands of type i; $[^iI_0] = [^iI] + [^iE_c]$	[mol/l]*
	Rate constants	
κ_i	Experimentally measured growth rate of type i in the exponential growth phase	$[s^{-1}]$
ρ_i	Experimentally measured relative rate of RNA synthesis per template strand of type i bound to replicase, including miscopying; $\rho_i = d[^iI_0]/([^iE_c]dt)$	$[s^{-1}]$*
^ik_A	Rate constant for replication complex formation in an RNA between strand of type i, with replicase; standard value 10^7	$[l/mol\,s^{-1}]$*
^ik_E	Rate constant for synthesizing and releasing a replica from a replication complex of type i, leaving behind an inactive complex; standard value 10^{-1}	$[s^{-1}]$
^ik_D	Rate constant for dissociation of the inactive complex of type i, dissociating into free RNA and replicase; standard value 10^{-2}	$[s^{-1}]$*
$^{ij}k_{ds}$	Rate constant for double strand formation between one strand of type i and the complementary strand of type j; for $i = j$ homoduplex, for $i \neq j$ heteroduplex formation; standard value 5×10^4	$[l/mol\,s^{-1}]$*
A_i	Relative rate of RNA synthesis, miscopying included, for a homogenous population of type i, leaving as template; $A_i = {^ik_E}[^iEI]/[^iI_0]$	$[s^{-1}]$
D_i	Relative template loss rate; mainly governed by double strand formation; $D_i = [^iI]/[^iI_0] \Sigma_j {^{ij}k_{ds}}[^jI]$	$[s^{-1}]$
E_i	Relative net excess RNA prodution rate with type i as template, including miscopying; $E_i = A_i - D_i$	$[s^{-1}]$*
$\bar{E}$	Relative next excess RNA production rate for all RNA types; $\bar{E} = \Sigma_i E_i x_i$	$[s^{-1}]$**
$\bar{E}_i, \bar{\bar{E}}$	$E_i, \bar{E}$ in the quasispecies steady state	$[s^{-1}]$*
*	Quantities that can be measured;	
**	quantities that are easy to measure.	

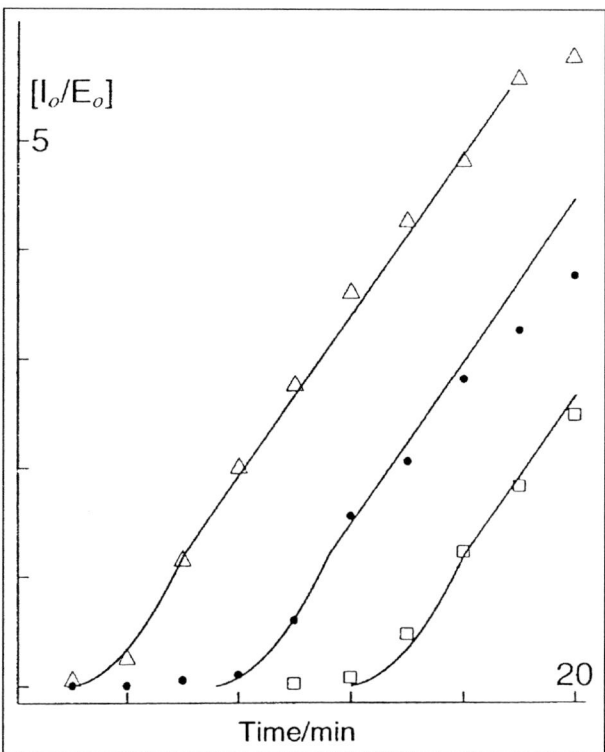

Fig. 5.1. Incorporation profiles of Qβ replicase. The mix contains buffer, Mg^{2+} ions and the four nucleoside 5'-triphosphates, one of them radioactively labeled. The acid-insoluble radioactivity is measured and the amount of RNA produced per enzyme molecule calculated. The overall replication rate in the linear growth phase, ρ, is determined from the slope of the linear part of the profile, $d\frac{[I_o]}{[E_o]}/dt$. The overall replication rate in the exponential growth phase, κ, is determined from recording the profile at different dilutions F_{dil} of the template RNA (in the above profiles 10^{-2} and 10^{-4}) and determining the rate from the displacement of the curve on the time axis Δt. $κ = ln F_{dil}/Δt$. From the profile shown here, $ρ = 6.1 \times 10^{-2} s^{-1}$ and $κ = 1.6 \times 10^{-2} s^{-1}$. See Biebricher, 1986 for more details.

The basic rate equations can be derived in a straightforward way. Let $[iI]$ be the molar concentration of unbound RNA species i. Free iI binds to replicase E with second-order rate constant ikA to form the active replication complex iEI, which synthesizes and liberates the replica RNA strand with first-order rate constant ikE. The remaining inactive complex iEI must dissociate into its components with first-order rate constant ikD before the enzyme can again serve as replicase. The template-enzyme complexes as well as the free complementary strands constitute replication intermediates analogous to different development groups or age groups of organisms; the sum $[iEc] = [iEI] + [iIE]$ represents the total concentration of replication intermediates, and the sum $[iI] + [iEI] + [iIE]$ corresponds to the total population of RNA species i, $[iI_o]$.

The rate equations are:

$d[^iEI]/dt = {}^ik_A[E][^iI] - {}^ik_E[^iEI]$

$d[^iIE]/dt = {}^ik_E[^iEI] - {}^ik_D[^iEI]$

Darwinian Evolution of RNA in Vitro

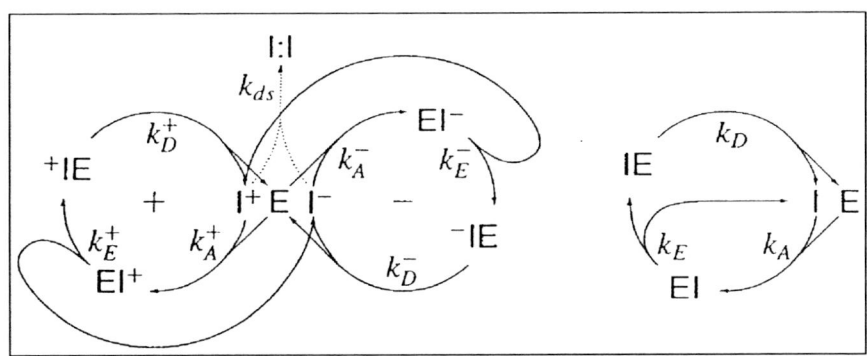

Fig. 5.2. Simplified mechanism of RNA replication by Qβ replicase. On the left, the simplified cross-catalytic cycle is shown, which can be further simplified to an autocatalytic one (on the right). RNA template (I) binds with the rate k_A to replicase (E) to form the active complex EI. A complementary strand is synthesized and released with the rate k_E. The remaining inactive complex IE dissociates with the rate k_D into its components. See Biebricher et al, 1983 for a more detailed mechanism.

$$d[^iE_c]/dt = {}^ik_A[E]\,[^iI] - {}^ik_D[^iIE] = -d[E]/dt$$

$$d[^iI]/dt = {}^ik_E[^iEI] + {}^ik_D[^iIE] = -{}^ik_A[E]\,[^iI]$$

$$d[^iI_o]/dt = {}^ik_E[^iEI] \tag{5.1}$$

It is often useful to describe the relative population change, and we obtain

$$A_i \frac{d[^iI_o]}{[^iI_o]dt} = {}^i\kappa_E \frac{[^iEI]}{[^iI_o]} \tag{5.2}$$

This equation, well known in population mathematics, states: The relative rate of population gain is equal to the proportion of the population in the reproducing age times its birth rate. We may define A_i as relative growth rate of species i; for the exponential growth phase $A_i = {}^ik$, where ^ik is the overall exponential growth constant of species i.

In animal populations, newborn individuals undergo long maturation periods before they reach reproductive age, while the time interval between deliveries in breeding adults is much shorter. In RNA reproduction, the opposite is the case: Nascent free replica strands bind rapidly to enzyme and immediately begin to reproduce, while the parent template requires a long period until it is recycled and ready to reproduce again (Biebricher et al, 1991).

How can we determine the proportion $[^iEI]/[^iI_o]$? It can not be measured directly, so indirect ways are needed to calculate it from quantities which can be measured. As shown initialy by Fisher, a growing population will have, after an equilibration period, a stable age distribution, leading to coherent growth of all intermediates:

$$A_i = \frac{d[^iI]}{[^iI]dt} = \frac{d[^iEI]}{[^iEI]dt} = \frac{d[^iIE]}{[^iIE]dt} = \frac{d[^iI_o]}{[^iI_o]dt} = {}^i\kappa \tag{5.3}$$

In the linear growth phase, after an equilibration period, a steady state is established in which the concentrations of the replication intermediates do not change and the flux through each step is equal to the reaction velocity v:

$$v = {}^ik_A[^iI]\,[E] = {}^ik_E[^iEI] = {}^ik_D[^iIE] = {}^i\rho[^iE_c] \tag{5.4}$$

Because the dissociation rate ^ik_D is rate-determining, and the concentration of free enzyme very small, $^i\rho[^iE_c] \approx {}^i\rho[^iE_o]$, where $[E_o]$ is the total replicase concentration, bound and free, both $^i\kappa$ and $^i\rho$ are thus functions of ^ik_A, ^ik_E and ^ik_D. At replicase concentrations in excess of 100 nM, template binding is found to be much faster than the two other steps. Under these conditions, $^i\kappa$ and $^i\rho$ are to a good approximation functions of ^ik_D and ^ik_E alone. $^i\kappa$ and $^i\rho$ can be readily determined as shown in Figure 5.1, and the ^ik_E and ^ik_D values can be calculated from them.

In the exponential growth phase, A_i is constant. In the linear growth phase, however, it varies because the fraction $[^iE_c]/[^iI_o]$ changes with time. We can define an apparent (steady state) enzyme binding constant

$$^iK_{IE} = \frac{[^iE_c]}{[E][^iI]} = \frac{^ik_A}{^i\rho} = \frac{^ik_A}{^ik_D}$$

that allows the calculation of the fraction $[^iE_c] / [^iI_o]$; note that $[^iI_o] = [^iE_c] + [^iI]$. In the linear growth phase, $[^iE_c] \approx [^iE_o]$ and thus $[^iI][E] = {}^iK_{IE}[E_o]$.

Double Strand Formation

At high concentrations of free RNA, the two complementary strands can react with one another to form double strands with the rate constant $^{ii}k_{ds}$. The double strands are found to be inactive as templates, and this reaction therefore leads to a loss of reproductive individuals (Biebricher et al, 1984, 1991). We assume in the following that double strands neither inhibit nor enhance replication of single-stranded RNA, an assumption which is not entirely realistic because double-stranded RNA does bind to enzyme and thus inhibits the replication process. Ignoring this effect, the relative population change due to destruction or death of individuals of type i is D_i, $D_i = 2\, ^{ii}k_{ds}[I^+][I^-]/[^iI_o]$ or—for equal concentrations of plus and minus strands—$D_i = 1/2 {}^{ii}k_{ds}[^iI]^2/[^iI_o]$. Destruction by other reactions, e.g., by hydrolysis, is negligible under normal experimental conditions.

Incorporation profiles (Fig. 5.1) do not show this destruction, and additional product analysis is required to measure the loss rate. In the exponential growth phase, newly-formed single strands are quickly complexed by replicase and thus protected against double strand formation. With most optimized RNA species, double strand formation in the exponential growth phase is not observed; however, during the replication of species having weaker secondary structures, replica and template may combine before the replica is released. A certain fraction of templates is thereby lost, a heavy disadvantage that favors selection of mutants where this does not happen. For optimized RNA species, we may neglect this reaction path.

The dynamic behavior can be summarized as follows: In the exponential growth phase, the dominant intermediates are bound to enzyme ($[^iE_c]$). In the linear growth phase, the enzyme is mostly complexed and free strands accumulate. With growing concentration of free strands, destruction by double strand formation becomes important, and soon a steady state is established where synthesis of new strands by replication is balanced by loss through double strand formation and the net relative production of RNA vanishes ($E_i... A_i - D_i \to 0$). In this growth phase, (denoted with a tilde), only the concentration of double strands increases, with $d([^iII])/dt = 1/2 {}^i\rho[E_o]$, and the steady state concentration of free strands is

$$[^i\tilde{I}] = \sqrt{2 {}_i[^iE_c]/{}^{ii}k_{ds}} \qquad (5.5)$$

Selection Among RNA Species

When two or more RNA species are present in the template population they compete with each other. As long as the sequences and the physical properties of the species differ, the outcome of the competition can be easily followed (Fig. 5.3). In this case, "type" is classified as all forms of the RNA strands belonging to the species, and mutation can be disregarded, because interconversion from one type to another is prohibited by the species barrier. Interpretation of the population dynamics is simplest in the exponential growth phase under conditions of small populations, when all resources required for amplification are present in excess. Under these conditions, each species grows with its characteristic rate; the population is enriched in the species with the higher growth rate and depleted in the species that grow more slowly. The fitness of each species is then characterized by its fecundity (Biebricher et al, 1985, 1991). The quantitative change of the population composition is determined by the growth constants:

$$\frac{[^1I_0]}{[^2I_0]_t} = \frac{[^1I_0]}{[^2I_0]_{t=0}} e^{(1k-2k)t} \tag{5.6}$$

The population composition changes exponentially, and very small differences in growth constants imply large composition changes at long times. When the κ differences are large, selection is dramatic: Suppose that an RNA species with a κ value 1/10 of that of an optimized species has to be amplified by a factor of 10. During that time a single strand of the optimized species is amplified by a factor of 10^{10}, i.e., to macroscopic appearance! When small populations containing several species are amplified, the slower species usually escape detection. For example, in an experiment where viral receptors were picked and sequenced (Domingo et al, 1978), only neutral or nearly neutral mutants were detected; seriously disadvantaged viruses will not have formed a visible plaque in the time period needed to amplify the faster ones to plaque size.*

For describing RNA evolution clear definitions and parameters that can be readily measured are required. Random effects may be neglected by assuming large populations. We define a selection value that describes the relative change of the relative population as follows:

$$S_i \equiv \frac{dx_i}{x_i dt} = {}^i\kappa \tag{5.7}$$

In the exponential phase, ς is constant and equal to the growth rate constant. In the linear growth phase, RNA selection is quite different: Often species with lower replication rate constants are selected (Fig. 5.3). In the linear growth phase, a resource—the replicase—is limiting, and the species that binds to a released replicase molecule most quickly will be selected, whatever its replication rate may be. Under these conditions, quantitative description of the selection process is more complicated because the selection value is a function of both type i and its competitors. The outcome can be determined most conveniently by computer simulations: The rate equations for competing species are set up to express sharing of resources. Calculated profiles again matched the experimental results with appropriate values taken for the rate constants. A typical example of a selection experiment, starting with two RNA species (MNV-11 and MDV-1) and its computer simulation is shown in Figure 5.4.

---

*What happens if both species grow at exactly the same rate? One would assume that then the composition does not change. However, as described in the previous Chapter, there are always growth fluctuations in finite populations. If no mechanism corrects the fluctuation, the population also fluctuates. Furthermore, in all selection events, the effects of fitness may be overcome by other events that work more or less at random. For example, in a serial transfer only a part of the population is taken for further breeding. If the aliquot and the population are small, individuals are selected mainly at random. If the contributions of fitness differences are smaller than those of such random events, a "neutral drift" could be observed.

Fig 5.3. Competition between the RNA species MNV-11 and SV-11 in the exponential (top) and linear (bottom) growth phases. The serial transfer experiment started with equimolar amounts of both species. Serial transfers in the exponential growth phase were diluted 1000-fold into fresh medium after incubation for 5 min at 30°C, in the linear growth phase diluted 50-fold after incubation for 30 min. The radioactively labeled species were separated by polyacrylamide gel electrophoresis and developed by autoradiography. The double band with the lowest mobility are hybrid partial double strands between MNV-11 and SV-11, the nucleotide sequences of which are strongly related one to another; the other bands are the single and double strands of MNV-11 and SV-11. Plus and minus strands of MNV-11 single strands are separated under these conditions. In the exponential growth phase, MNV-11 is selected due to its higher overall growth rate, while SV-11 is selected in the linear growth phase due to its faster replicase binding rate. After Biebricher and Luce, 1992.

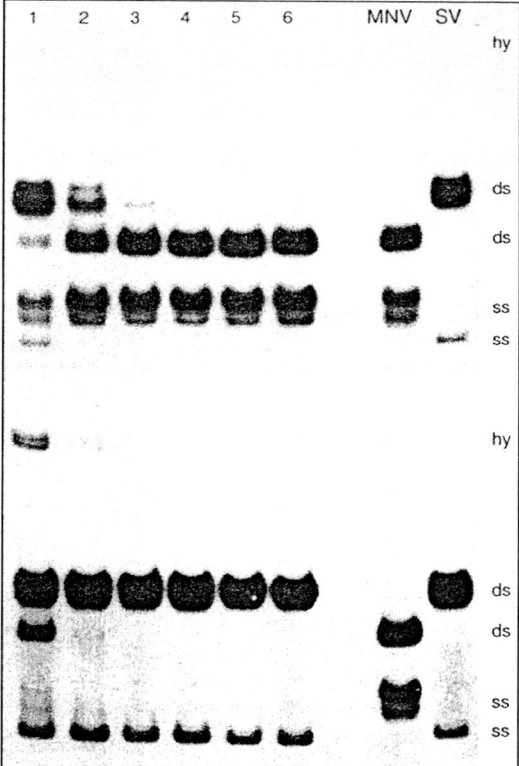

In the late linear growth phase, loss by double strand formation must also be taken into account in modeling the selection process. When the RNA sequences differ significantly, the formation of heteroduplexes is negligible. Species with low concentrations of free single strands are favored by low loss rates through double strand formation; eventually the population will reach a steady state where its relative composition, aside from the growing double strand concentration, will not change anymore: A stable ecosystem has been formed (Fig. 5.4). The value of the model system is illustrated by the fact that the homologous phenomena observed in organismic evolution—selection by fecundity, by competition for resources or by death rate reduction—not only occur in this system, but can be clearly quantified (Biebricher et al, 1985, 1991).

Mutation in Replicating RNA

For describing the interplay of mutation and selection we have to define as type i the mutants of an RNA species. As noted in the previous Chapters, the experimental challenge is great, because the types have to be distinguished by their genotypes and because the only rigorous way of determining the mutation rate is to avoid selection by restricting growth to a single generation. In RNA synthesis, this is rarely possible, and we thus must nearly always consider contributions both from mutation and selection. From in vitro (Batschelet et al, 1976) and in vivo studies (Drake, 1993; Drake and Holland, 1999), average error rates per nucleotide incorporated are estimated to be of the order of 10^{-4}. This implies that while the majority of phage RNA progeny are mutants, for the much shorter sequences used in in vitro experiments the majority of the replicas should have the same, i.e., correct, sequence.

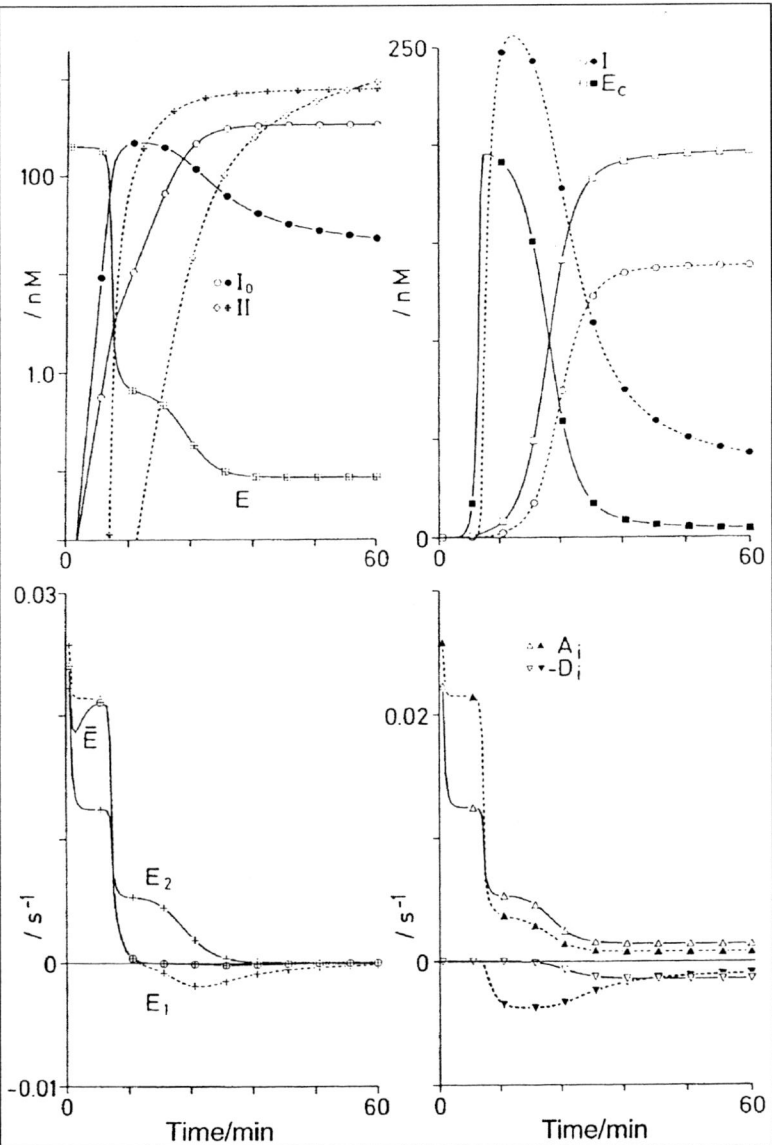

Fig. 5.4. Simulation of the competition between two RNA species. Standard rate parameters for the species were used, except for $^2k_A = 4 \times$ standard and $^2k_D = 1/4$ standard values. $^{11}k_{ds} = {}^{22}k_{ds} = 5 \times 10^4 {}_s^{-1}$. Initially both species are present at concentrations of 1 pM. In the exponential growth phase (0-8 min), the smaller ^2k_D value is detrimental: species 1 (filled symbols) grows more rapidly to saturate the enzyme and enter a steady state of double strand formation (8-12 min); its net synthesis $\bar{E}_1$ vanishes. Species 2 (open symbols) continues to grow exponentially with a lower rate. Macroscopic selection in the linear growth phase takes place when species conquers the majority of the enzyme due to its higher enzyme binding rate ^1k_A (12-40 min). Eventually, an ecosystem is formed where E_1, E_2 and $\bar{E}$ disappear. After Biebricher et al, 1985.

The experimentally determined mutant spectrum of the replicating RNA species MNV-11 (Rohde et al, 1995), however, was surprisingly different: After many serial transfer passages under constant conditions to reach population equilibrium, a broad mutant spectrum was found (Fig. 5.5). Mutant spectra of populations produced by growth under slightly different conditions were significantly different fom one another, and the implications of the quasispecies concept described in the previous Chapter were fully confirmed. For the analysis of the sequence and the properties of the mutants, a representative collection of mutants was cloned under conditions where the mutant spectrum was not distorted by selection bias and error multiplication. This was achieved by cloning the RNA into DNA and sequencing a representative number of the cDNA clones. The RNA genotype that initiated the clone could be readily (and accurately!) reproduced by transcription from the DNA clones. This may seem implausible because the fidelities of transcription and replication are about equal. However, transcription uses exclusively the original (DNA) template, thus avoiding error propagation, while replication also uses the copy as template.

The mutant spectrum of MNV-11 (Fig. 5.5) was found to be broad. The master sequence was the most frequent genome, but it represented always a minority of the quasispecies population. Some positions of the sequence were highly conserved while others varied. Base transitions, base transversions, base deletions and base insertions were found, in one case a duplication of a 7 base motif. Mutations are independent events, and if mutations were predominantly responsible for the mutant frequency landscape, one would expect to find a high frequency of one-error-mutants and much smaller frequencies of two-error- or multi-error mutants. This was not the case, and mutants with up to 10% of the positions altered were observed. Therefore, the mutant frequency landscape is predominantly shaped by the selection values, and indeed the analysis of isolated mutants showed that they are neutral or nearly-neutral. In the linear growth phase, selection is dependent on many parameters, and the master was the best compromise in replication speed, competing for replicase and minimizing double strand formation.

Experiments suggested that RNA structural elements are crucial for maintaining the replication efficiency (Zamora et al, 1995). Mutations that disturbed the secondary structure could often be compensated by further mutations restoring their replication efficiency. Directed mutations of the Qβ genome were also shown to result in compensating pseudorevertants (Olsthoorn and van Duin, 1996; Klovins et al, 1997).

Of particular interest is evolution in the exponential growth phase, because the selection values are then equal to the overall replication rate constants. Therefore, there are fewer constraints than in the linear growth phase, and as a consequence, the master sequence was found to be degenerate (Fig. 5.5). Remarkably, the master sequence of the linear growth phase was not detectable in these experiments.

Adaptation of RNA in Vitro

When conditions for replication of MNV-11 were changed, e.g., by increasing the ionic strength or adding replication inhibitors, a rapid replacement of the mutant spectrum by a new one was observed (Fig. 5.5). In agreement with the theoretical results of the quasispecies, the first step in evolution is the selection of the most adapted mutants already present in the quasispecies, and not—as previously assumed—the generation of new mutants. When the concentration of the adapted mutants rises, new mutations are formed with a higher probability, and the mutant frequency landscape floats in the sequence plane to the positions dictated by the fitness landscape. Eventually, a new stable equilibrium mutant spectrum is built up around a new master sequence.

The adaptation of RNA species to new phenotypes has been the subject of many qualitative and quantitative studies. The speed of adaptation is strongly dependent on the population size. In small populations, many steps are required to reach a new equilibrium, and each of them

Darwinian Evolution of RNA in Vitro

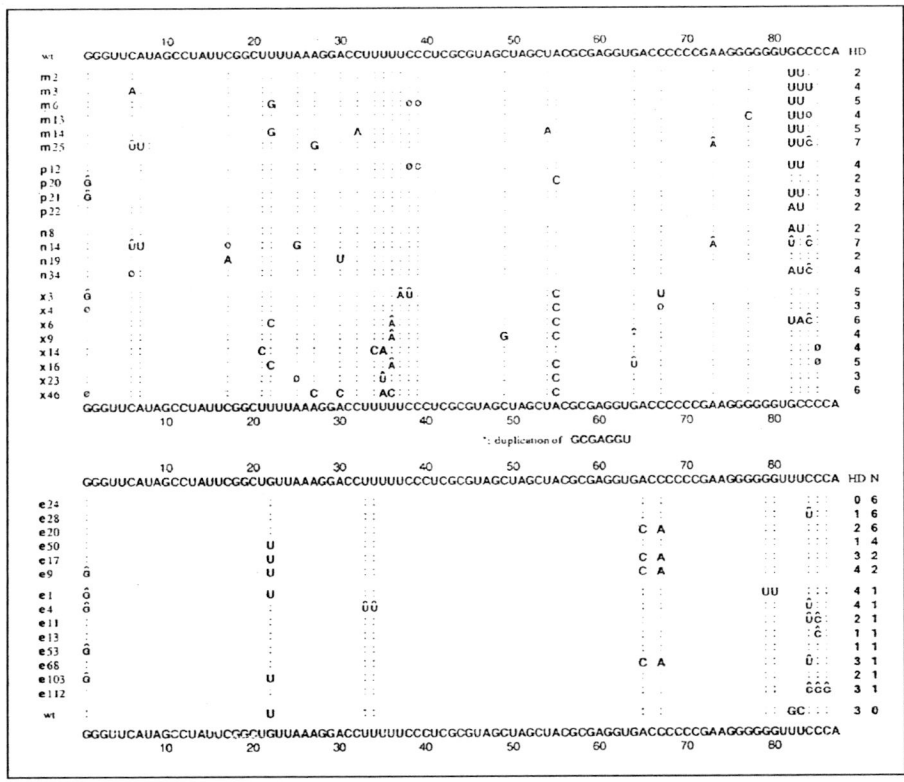

Fig. 5.5. Sequence determination of representative subclones of quasispecies populations of MNV-11 grown in the linear growth phase (upper part) and exponential growth phase (lower part). Populations are indicated by letters: m, 5 h incubation in the linear growth phase; n,p, after 2 h incubation in the linear growth phase the strands are melted and separated into plus (p) and minus (n) strands; x, 8 h. Incubation in the linear growth phase in the presence of 50 mM $(NH_4)_2SO_4$. Numbers indicate the mutant isolate. All mutants occurred only once, the wild type (wt) occupied about 40% of the population. The number in the last column is the number of deviations from the wild type sequence. The population 1 was obtained by 10^{30}-fold amplification in the exponential growth (10 serial transfers with dilution factors of 1000 and incubation of 6 min). The master sequence was degenerated (N in the last columns indicates the absolute mutant frequency). The wild type of the linear growth type was not found in this population. After Rohde et al, 1995.

must have a small selective advantage of its own. Small populations and "rugged" fitness landscapes usually result in irreproducible pathways through the sequence plane and different outcomes of evolution experiments under otherwise identical conditions (Biebricher, 1983). The chance to populate a position with a high fitness in a Hamming distance of a few steps depends on the Hamming distance itself, the number of possible routes to reach it, and on the the deepness of the ditches that have to be crossed. If one of the steps leads to a lethal mutant that can not replicate, this position in sequence space must be crossed with a jump, i.e., by simultaneous mutations in more than one position.

The first quantitative adaptation experiment reported was the adaptation of the species MDV-1 to small concentrations of ethidium bromide (Kramer et al, 1974). The adapted sequence (averaged over the RNA population) was altered at three positions, which had to be selected step by step, because the transfers started with a population of only 10^6 strands. The single-error mutant was already in the quasispecies population, and the next mutations occurred in the 7th and 12th transfer, respectively. As expected, the adapted mutant was slightly inferior in the absence of ethidium bromide. However, the selection was difficult to interpret, because of the small populations and the strongly changing selection pressure when going through the different growth phases.

Eigen and collaborators (Schober et al, 1995) have developed a machine that avoids these disadvantages. It keeps the species in the exponential growth phase, because real time RNA concentration measurements trigger the next serial transfer before the enzyme is saturated. Furthermore, RNA populations never drop below 10^{11}. Using this device, a variant MNV-11 resistant against RNaseA has been selected. The sequence of the final product has been determined, but the route through sequence space is unknown.

Recombination Among RNA Molecules

In organisms with large DNA genomes the high DNA replication fidelity makes large jumps impossible and adaptation via mutations extremely slow. Adaptation involving larger changes of the genotype is only possible via the alternative route of recombination. Recombination in most RNA viruses has been documented only over the last two decades because it is often rare and thus difficult to detect (Lai, 1993; Palasingam and Shaklee, 1992; Weiss and Schlesinger, 1991). Several alternative mechanisms of recombination have been proposed, the simplest and most plausible being "copy choice", i.e., the replicase switches templates during a replication round. RNA recombination is usually only detected when an advantageous product is formed, e.g., by restoring wild type from two different, seriously handicapped mutants (see also Chapter 6).

In the Qβ-replication system an RNA species replicated by Qβ replicase was isolated which appears to be a recombinant between part of the replicase gene of Qβ and a tRNA (Munishkin et al, 1988). RNA recombination by Qβ replicase in vitro has also been observed (Biebricher and Luce, 1992). In the evolutionary optimization of RNA species in vitro, RNA duplications produced by RNA recombination, followed by mutation, seem to be an important pathway to increase sequence complexity (Biebricher, 1983; Biebricher and Luce, 1992, 1993). Due to the rarity of RNA recombination in some viruses, its importance in virus evolution is difficult to assess, but it may have been underestimated for a number of viruses (Chapter 6).

Can Biological Information Be Generated De Novo?

All experiments described so far showed Darwinian adaptation to the environment, i.e., optimization of a preexisting biological function. Evolution, however, is able not only to adapt but also to create. Is it possible to obtain a replicating RNA in the absence of a template?

Such a proposal might seem tantamount to a relapse into old, erroneous theories of de novo synthesis of vermin organisms from organic waste. On the molecular level, the circumstances are different: Recently many experimenters succeeded in selecting RNA with entirely new functions from random RNA sequences (Chapter 8). Nobody is troubled by these experiments, because the human ingenuity behind them is recognized as a potent driving force. Natural selection can do the same, however, even though the probability of finding a useful locus in the vast sequence space is extremely low. On the other hand, we have seen in the previous Chapter that it is not necessary to hit a particular locus: It suffices to hit a mountainous region in the fitness landscape. The route to the peak(s) is then guided by evolutionary forces. In other words, the large number of total losers in the lottery is compensated in part by the large number of minor wins.

Indeed, it has been possible to create novel replicable RNA species. Two basic strategies have proved successful:

1. In one procedure winners are selected from a huge library containing randomly assembled sequences (Biebricher and Orgel, 1973; Brown and Gold, 1993).
2. In the other, very pure RNA replicase at high concentrations is incubated with high nucleotide triphosphate concentrations. After long incubation periods, replicable RNA was produced (Sumper and Luce, 1975). As one would expect from such experiments, different RNA species were selected in each experiment (Biebricher et al, 1981a; Biebricher, 1988), even if they were produced under completely identical conditions (Fig. 5.6).

The features of this remarkable reaction have been described (Biebricher et al, 1981a,b, 1986, 1993; Biebricher, 1987; Biebricher and Luce, 1993). The reaction kinetics is fundamentally different from template-instructed replication (Biebricher et al, 1981b); it requires special conditions to take place. Initially, nucleoside triphosphates are condensed at a rate five orders of magnitude more slowly than template-instructed incorporation in a more or less random manner (Biebricher et al, 1986) to oligo- or polynucleotides with chain lengths from 5-50. The first replicable RNA products that can be isolated have short chain lengths (25-40 nucleotides) and are rather inefficient templates (Biebricher and Luce, 1993). Their primary sequences are not related to one another, but their secondary structures are (Fig. 5.6) (Biebricher and Luce, 1993). The reaction is suppressed by the presence of nucleic acids; when these are present, the only RNA produced is derived from the input RNA (Avota et al, 1997). It is remarkable that the de novo generation of replicable species succeeds even with DNA-dependent RNA polymerases that have no natural RNA templates. RNA templates replicated by T7 RNA polymerases (Biebricher and Luce, 1996) and by RNA polymerase from *E. coli* (Biebricher and Orgel, 1973; Wettich, 1999) have been isolated and characterized. The mechanism of RNA amplification by these enzymes is very similar to that described for Qβ replicase. The templates are strictly specific to their amplifying enzymes, i.e., a template replicated by Qβ replicase can not by amplified by RNA polymerase from T7 or *E. coli* and vice versa.

The first inefficient products of template-free synthesis rapidly undergo an (irreproducible) optimization process by recombination and mutation, thereby enhancing their sequence complexity.

Conditions for Replication

The fitness phenotype of a replicating RNA is its efficiency to direct the replicase to amplify it. It would be highly desirable if the phenotype could be directly derived from the genotype, or if one could design RNA genotypes with suitable phenotypic properties. A possible approach to solve this problem is to compare the genotypes of the large number of replicating species that have been isolated. The most conspicuous properties are the invariant ends: at the 5' terminus pppGG(G), at the 3' end CCA. All tRNA sequences have the same end; however, enzymes tested that recognise tRNAs, e.g., tRNA CCA pyrophosphorylase or an amino acyl tRNA synthetase, do not accept Qβ templates (Biebricher, unpublished). The invariant ends are a necessary property but not a sufficient one for replication because many RNA sequences with this sequence requirement turn out not to be accepted by replicase.

Visual comparison and sequence alignment programs do not reveal further consensus motifs in the primary sequence. However, comparison of the secondary structures indeed showed a striking coincidence: the 5' termini of the calculated structures were involved in a double-helical stem, while their 3' termini were unstructured (Fig. 5.7) (Biebricher and Luce, 1993). There is evidence that these structural features also apply for apparent exceptions (Zamora et al, 1995). In one example (species SV-11), the genotype is only replicable in a metastable structure. When the metastable structure is converted into the stable one, template activity is lost (Biebricher and Luce, 1992). The constraints to meet the structural requirements are stricter than one would think, because they apply to both complementary strands. Were the base-pairing restricted

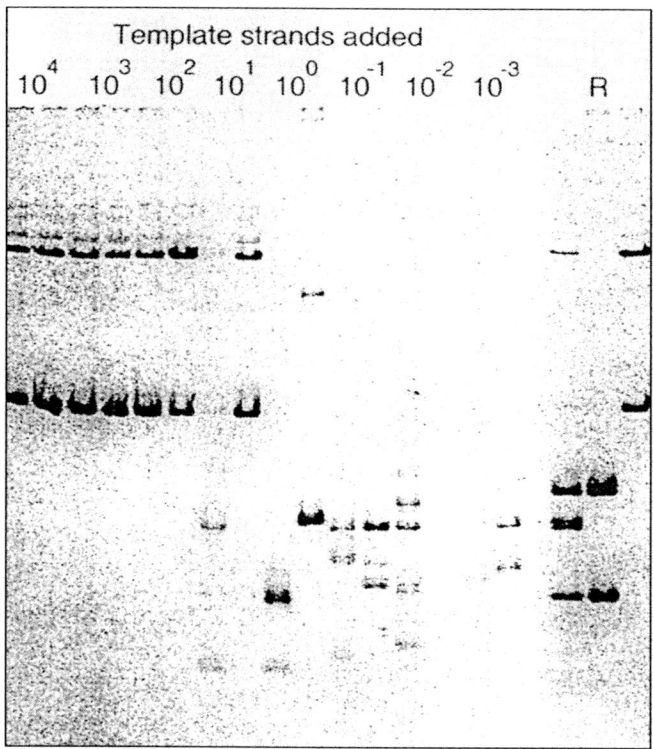

Fig. 5.6. Template-instructed and template-free synthesis by Qβ replicase. species MDV-1 was serially diluted and the number of strands indicated were used to inoculate an RNA replicase mixture containing 140 nM Qβ replicase and incubated for 16 h at 30°C. The average number of strands required to trigger template-directed growth was reproducibly found to be about 20; the reason is still unknown. At lower concentrations or without addition of template, small and irreproducible products were detected after long lag times. (After Biebricher, 1987).

to the Watson-Crick type, a stem in the 5' terminus of one strand would lead to a 3' terminal stem in the complementary strand. Noncanonical base-pairing and helix imperfections are found in the sequences to obtain the desired structure. Converting these to the canonical base pairs destroys the template activity (Zamora et al, 1995). Design and synthesis of RNA strands that meet this criterion resulted in replicable species, although the replication efficiency was low and was improved by evolutionary adaptation during the experiment. Additional features are probably required to improve the efficiency. Pyrimidine clusters improve binding (Brown and Gold, 1996) but binding and replicability are poorly correlated. The helix imperfections required by the criterion lead to the existence of suboptimally folded structures with similar structural energies. It is conceivable that the RNA undergoes a structural change when binding to replicase. However, dynamic effects are very difficult to study. A partial exploration of the "shape space" (Schuster et al, 1994) for template activity could bring further insight.

Ongoing research with many RNA viruses is providing a picture of mutation, competition and selection in fitness landscapes which is conceptually remarkably close to the view derived with simple RNA replicons. In the next Chapters these processes of movement of RNA viruses through sequence space and fitness landscape are reviewed.

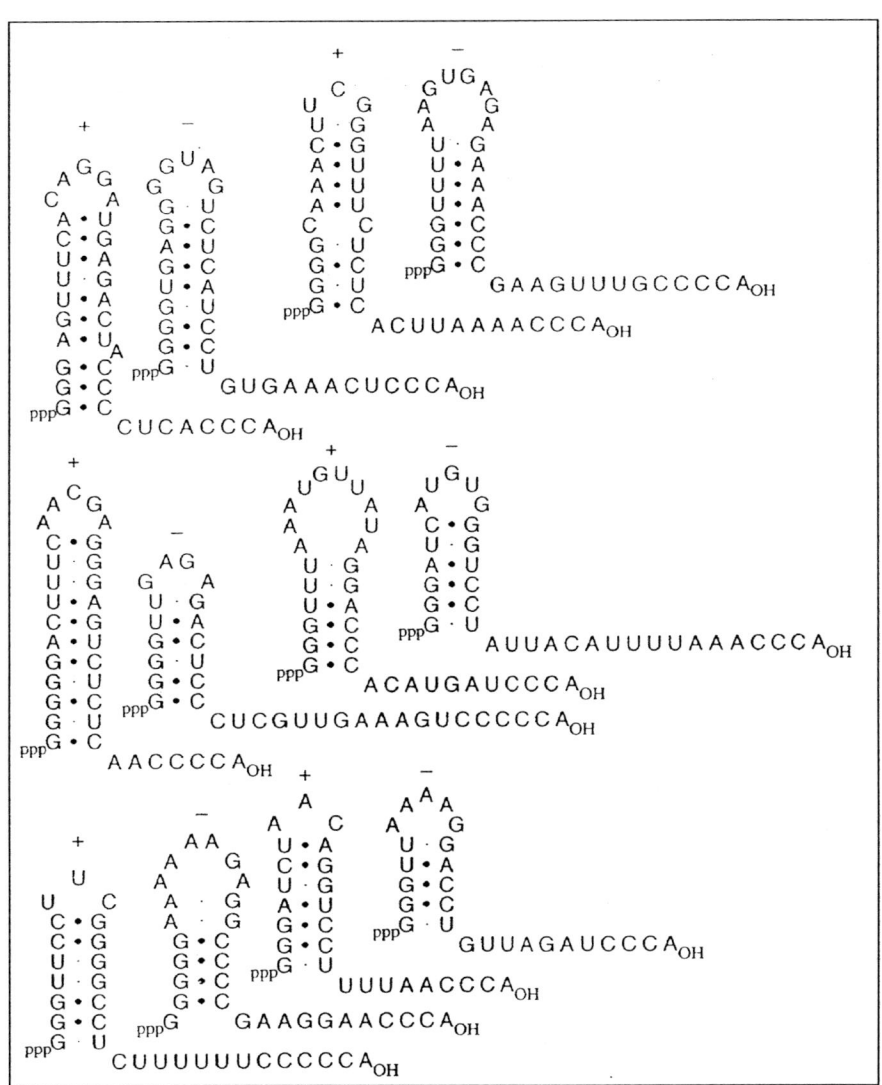

Fig. 5.7. Primary and (tentative) secondary structures of short-chained replicable RNA species. Note that the 5' termini are involved in a stem structure while the 3' termini are not. See Biebricher and Luce, 1993 and Zamora et al, 1995 for more details.

References

1. Avota E, Berzins V, Grens E et al. The natural 6S RNA found in Qβ-infected cells is derived from host and phage RNA. J Mol Biol 1998; 276:7-17.
2. Batschelet E, Domingo E, Weissmann C. The proportion of revertant and mutant phage in a growing population, as a function of mutation and growth rate. Gene 1976; 1:27-32.
3. Biebricher CK. Darwinian selection of self-replicating RNA. Evol Biol 1983; 16:1-52.
4. Biebricher CK. Darwinian evolution of self-replicating RNA. Chemica Scripts 1986; 26B:51-57.
5. Biebricher CK. Replication and evolution of short-chained RNA species replicated by Qβ replicase; Cold Spring Harb. Symp Quant Biol 1988; 52:299-306.

6. Biebricher CK, Luce R. In vitro recombination and terminal elongation of RNA by Qβ replicase. EMBO J 1992; 11:5129-35.
7. Biebricher CK, Luce R. Sequence analysis of RNA species synthesized without template. Biochemistry 1993; 32:4848-54.
8. Biebricher CK, Luce R. Template-free synthesis of RNA species replicating with T7RNA polymerase. EMBO J 1996; 15:3458-65.
9. Biebricher CK, Orgel LE. An RNA that multiplies indefinitely with DNA-dependent RNA polymerase: Selection from a random copolymer. Proc Natl Acad Sci USA 1973; 70:934-938.
10. Biebricher CK, Eigen M, Luce R. Product analysis of RNA generated de novo by Qβ replicase. J Mol Biol 1981a; 148:369-390.
11. Biebricher CK, Eigen M, Luce R. Kinetic analysis of template-instructed and de novo RNA synthesis by Qβ replicase. J Mol Biol 1981b; 148:391-410.
12. Biebricher CK, Eigen M, Gardiner WC. Kinetics of RNA replication. Biochemistry 1983; 22:2544-59.
13. Biebricher CK, Eigen M, Gardiner WC. Kinetics of RNA replication: Plus-minus asymmetry and double-strand formation. Biochemistry 1984; 23:3186-94.
14. Biebricher CK, Eigen M, Gardiner WC. Kinetics of RNA replication: Competition and selection among self-replicating RNA species. Biochemistry 1984; 24:6550-60.
15. Biebricher CK, Eigen M, Gardiner WC. Quantitative analysis of selection and mutation in self-replicating RNA. In: Biologically Inspired Physics (Peliti L, Hrsg.) NATO ASI Series B. New York: Plenum Press 1991; 263:317-337.
16. Biebricher CK, Eigen M, Luce R. Template-free RNA synthesis by Qβ replicase. Nature 1986; 321:89-91.
17. Brown D, Gold L. Selection and characterization of RNAs replicated by Qβ replicase. Biochemistry 1995; 34:14775-782.
18. Domingo E, Sabo DL, Tanigulli T et al. Nucleotide sequence heterogeneity of an RNA phage population. Cell 1978; 13:735-44.
19. Drake JW. Rates of spontaneous mutation among RNA viruses. Proc Natl Acad Sci USA 1993; 90:4171-75.
20. Eigen M, Biebricher CK, Gebinoga M et al. The Hypercycle. Coupling of RNA and protein biosynthesis in the infection cycle of an RNA bacteriophage. Biochemistry 1991; 30:11005-18.
21. Klovins J, Tsareva NA, de Smith MH et al. Rapid evolution of translational control mechanisms in RNA genomes. J Mol Biol 1997; 265:372-84.
22. Kramer FR, Mills DR, Cole PE et al. Evolution in vitro: Sequence and phenotype of a mutant resistant to eithidium bromide. J Mol Biol 1974; 89:719-36.
23. Lai MMC. Genetic recombination in RNA viruses. Curr Top Microbiol Immunol 1992; 176:21-32.
24. Levisohn R, Spiegelman S. The cloning of a self-replicating RNA molecule. Proc Natl Acad Sci USA 1968; 60:866-872.
25. Levisohn R, Spiegelman S. Further extracellular Darwinian experiments with replicating RNA molecules: diverse variants isolated under different selective conditions. Proc Natl Acad Sci USA 1969; 63:807-811.
26. Mills DR, Kramer FR, Spiegelman S. Complete nucleotide sequence of a replicating RNA molecule. Science 1973; 180:916-927.
27. Mills DR, Peterson RL, Spiegelman S. An extracellular Darwinian experiment with a self-duplicating nucleic acid molecule. Proc Natl Acad Sci USA 1967; 58:217-224.
28. Munishkin AV, Voronin LA, Chetverin AB. An in vivo recombinant RNA capable of autocatalytic synthes by Qβ replicase. Nature 1988; 333:473-475.
29. Olsthoorn RCL, van Duin J. Random removal of inserts from an RNA genome-selection against single-stranded RNA. J Virol 1996; 70:729-736.
30. Orgel LE. Selection in vitro. Proc R Soc Lond 1979; B205:435-442.
31. Palasingam K, Shaklee PN. Reversion of Qβ RNA phage mutants by homologous RNA recombination. J Virol 1992; 66:2435-42.
32. Rohde N, Daum H, Biebricher CK. The mutant distribution of an RNA species replicated by Qβ replicase. J Mol Biol 1995; 249:754-762.

33. Saffhill R, Schneider-Bernloehr H, Orgel LE et al. In vitro selection of bacteriophage Qβ variants resistant to ethidium bromide. J Mol Biol 1970; 51:531-539.
34. Schaffner W, Rüegg KJ, Weissmann C. Nanovariant RNAs: Nucleotide sequence and interaction with bacteriophage Qβ replicase. J Mol Biol 1977; 117:877-907.
35. Schober A, Walzer NG, Tangen U et al. Multichannel PCR and serial transfer machine as structure tools in evolutionary biotechnology. Biotechniques 1995; 18:652-660.
36. Schuster P, Fontana W, Stadler PF et al. From sequences to shapes and back: A case study in RNA secondary structures. Proc Roy Soc London B 1994; 255:279-284.
37. Spiegelman S. An approach to the experimental analysis of precellular evolution. Quart Rev Biophys 1971; 4:213-253.
38. Spiegelman S, Haruna I, Holland IB et al. The synthesis of a self-propagating and infectious nucleic acid with a purified enzyme. Proc Natl Acad Sci USA 1965; 54:919-927.
39. Sumper M, Luce R. Evidence for de novo production of self-replicating and environmentally adapted RNA structures by bacteriophage Qβ replicase. Proc Natl Acad Sci USA 1975; 72:162-166.
40. Weiss BG, Schlesinger S. Recombination between sindbis virus RNAs. J Virol 1991; 65:4017-25.
41. Wettich A. In vitro RNA replication with DNA-dependent RNA polymerase from *E. coli*. Göttingen: Hainholz Verlag, 1999.
42. Zamora H, Luce R, Biebricher CK. Design of aritificial short-chained RNA species that are replicated by Qβ replicase. Biochemistry 1995; 34:1261-66.

CHAPTER 6

Experimental Studies on Viral Quasispecies

In contrast to replication of simple, noninfectious RNA molecules in vitro analyzed in the previous Chapter, multiplication of infectious virus necessitates a concatenation of steps from entry into a cell until the release of progeny particles (Chapter 2). Each of the specific steps in the virus life cycle may be influenced by the occurrence of mutations, competition, and selection for more efficient performance. In this Chapter we examine mechanisms of genetic variation of RNA viruses and the quasispecies dynamics in RNA virus populations.

Molecular Mechanisms of Genetic Variation of RNA Viruses: Mutation, Recombination and Genome Segment Reassortment

Mutations, Replication Rounds, and Their Effects on Viral Quasispecies

RNA viruses use all mechanisms of genetic variation known to operate in cellular DNA: mutation, homologous and nonhomologous recombination and genome segment reassortment (Fig. 6.1). Mutation rates during replication and retrotranscription of RNA genetic elements have been estimated in 10^{-4} to 10^{-5} misincorporations per nucleotide copied. These values are approximately 10^4-to 10^6-times higher than those normally operating during cellular DNA replication (Eigen and Biebricher, 1988; Drake, 1991, 1993; Domingo and Holland, 1994; Drake et al, 1998; Drake and Holland, 1999) (Fig. 6.2.). Mutation rate refers to the occurrence of mutations during a single round of template copying. Therefore it describes a biochemical event independent of the competitive ability of the parental genome and of the mutant genomes that are produced. Mutation rate must be distinguished from the mutation frequency which is the proportion of mutations in a genome population. The term mutant frequency is sometimes used to refer to the frequency of a specific type of mutant (resistant to an antibody, drug, etc.) in a population of viral genomes (Table 6.1). Mutation frequency is a population number which will be influenced by the competitive replication ability of the mutant genomes relative to the parental genomes. A mutant may be generated with a high rate, but its frequency may soon become very low if the mutation decreases the relative replicative capacity of the molecule harboring it (Fig.6.3). A significant example is provided by measles virus, which shows a considerable antigenic and genetic stability in the field. Yet it has been estimated that it mutates at a rate of 9×10^{-5} substitutions per base copied (Schrag et al, 1999), a value which is within the range displayed by highly variable RNA viruses (Drake and Holland, 1999). Long-term antigenic stability is not necessarily related to mutation frequencies at the antigenic sites of viruses. Among the family of *Picornaviridae*, the cardioviruses, poliovirus, foot-and-mouth disease virus, and rhinovirus comprise 1, 3, 7, and more than 100 serotypes, respectively. In spite of these logarithmic differences among serotypes, the frequencies of monoclonal antibody-resistant mutants were in the range of 10^{-3} to 10^{-5} in all cases, without any trend towards higher frequencies for rhinoviruses (Domingo and Holland, 1994).

Quasispecies and RNA Virus Evolution: Principles and Consequences, by Esteban Domingo, Christof K. Biebricher, Manfred Eigen and John J. Holland. ©2001 Eurekah.com.

Experimental Studies on Viral Quasispecies 83

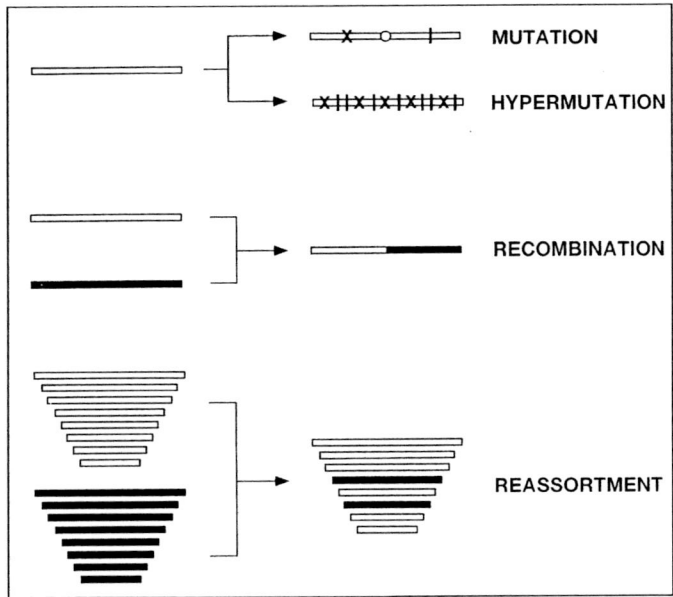

Fig. 6.1. Schematic representation of the process of mutation, hypermutation, recombination, and genome segment reassortment. Genomes are represented by empty or filled thin bars. Symbols on the bars represent mutations. Hypermutated genomes often include a dominance of some specific types of mutations.

In 1979, an avian influenza H1N1 virus crossed the species barrier and infected pigs showing higher evolutionary rates than classic swine or human influenza viruses. Although initially the occurrence of a "mutator" mutation in the newly established swine influenza virus was postulated, the estimates of amantadine-resistant mutants of a H2N2 and the H1N1 did not provide evidence for the presence of the "mutator" mutation (Stech et al, 1999).

It must be stressed that mutation rates are known to vary with a number of environmental influences, and that, in consequence, they cannot be taken as absolute values. For example, even for a given template, a specific position, and a defined copying enzyme, the misincorporation rate may vary with metal ion concentration, and with the relative concentrations of the four nucleotide substrates. Intracellular substrate concentrations may be dependent on the cell types, and their metabolic state. The effects of substrate concentrations on misinsertion frequencies have been documented in cell culture and also in cell-free systems, with purified viral replicases and reverse transcriptases (Meyerhans and Vartanian, 1999). Human immunodeficiency virus type 1 (HIV-1) mutation rates in vivo are about one order of magnitude lower than estimated with purified reverse transcriptase in vitro. This difference may partly be due to the presence of Vpr (a regulatory protein encoded by HIV-1) in virus particles (Mansky, 1996). Even assuming identical environmental conditions in all respects imaginable, there will be a quantum mechanical uncertainty regarding when a misincorporation event might occur. In spite of the indeterminacy concerning precise mutation rate values at each genomic nucleotide and the stochastic nature of mutation occurrence, the continuous generation of virus mutants in cell culture and in infected organisms has been amply documented, and it is currently regarded as a key adaptive strategy of RNA viruses.

With an average RNA virus mutation rate of approximately 10^{-4} substitutions per nucleotide copied, the progeny viral RNA (or cDNA) synthesized in an infected cell will contain on

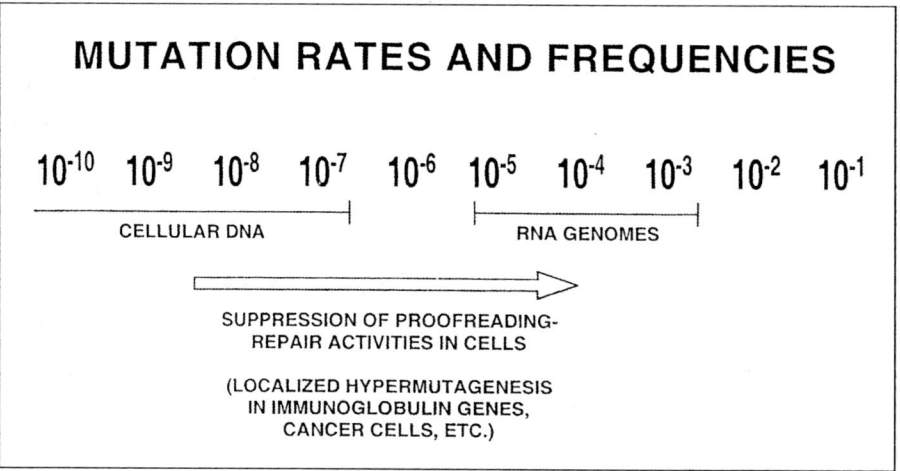

Fig. 6.2. Mutation rates and frequencies for RNA and DNA genomes. The range of values have been estimated by genetic and biochemical methods which are summarized in the text. Cellular DNA may display high mutation rates induced by a variety of mechanisms which are not well understood (arrow).

average approximately one mutation per RNA (or cDNA) molecule. Therefore, most progeny genomes will differ from their parental templates. If the mutation rate were one order of magnitude lower, approximately ten rounds of genome replication would be needed to produce on average a genomic molecule with one mutation. A typical infection of one BHK-21 cell with one infectious foot-and-mouth disease virus (FMDV) particle yields 2×10^6 physical particles (of which about 200 are infectious). Since more than 50% of the physical particles contain one genomic RNA molecule, this means that the viral genome underwent an equivalent of 21 doublings in a single cell (although it is not known how many of the intermediate minus strands—which are a minority with respect to plus strands—are active as templates for plus strand genomic RNA synthesis). Typically, in a standard cell culture experiment, 10^6 cells are infected with FMDV, and in experiments intended to reach high population sizes, up to 10^8 cells have been infected. This means that 2×10^9 rounds of genome replication (and mutation testing!) occur in a single cell culture bottle infected at high multiplicity of infection with FMDV. Of course, for some viruses, genome multiplication is not so exuberant. It has been estimated that cells infected with Borna disease virus—a virus first identified as a disease agent of horses at the beginning of the 20th century, and later found to infect a number of animal hosts—only about 0.5 infectious units can be detected from each cell. This may reflect low numbers of replication rounds, and may contribute to the limited genetic diversity among independent Borna isolates, in spite of the virus encoding a polymerase which is related to that of other negative strand RNA viruses.

For other viruses, the multiplication potential is even orders of magnitude larger than for FMDV. This is the case for bacteriophage Qβ or vesicular stomatitis virus (VSV) with titers of up 10^{11} or 10^{12} infectious units per ml, attained in standard cell culture infections. Yet if sequential samples of genomic RNA of FMDV, Qβ or VSV are analyzed, the number of mutations seen in their consensus genomic sequences is modest. Why? What prevents mutations from accumulating? This is a key issue since the same question transferred to the in vivo replication of viruses bears on the maintenance of virus identity through rounds of natural infections, transmissions, etc.

Table 6.1. Mutation rate and mutation frequency[a]

Definition of These Concepts as Used in Virology

MUTATION RATE is the frequency of occurrence of a mutation event during genome replication. It is a biochemical event dependent on a number of factors (the nature of the enzyme involved in replication, template, ionic environment, temperature, etc.). It is *independent* of the relative fitness of the genomes with or without mutations introduced as a result of such biochemical event. (Fitness describes the adaptation of an organism to its environment. For viruses, relative fitness values are quantitated as their capacity to produce infectious progeny relative to a reference virus in a defined environment; see Chapter 7.)
MUTATION FREQUENCY is the proportion of mutations relative to the consensus sequence found in a genome population. If applied to a specific mutation it is sometimes termed *mutant* frequency. It is a population number, *dependent* on the relative fitness of the genomes harboring the different mutations.
HOT SPOT is a sequence position with a particularly high mutation frequency.

[a]An example which illustrates the difference between mutation rate and mutation frequency is depicted in Figure 6.3.

The answer to this question relates to the very essence of the quasispecies structure of RNA viruses as discussed in preceding Chapters. Once generated, mutants are subjected to a process of competition for replication. In a cell culture dish this can be easily visualized since those viruses which first complete their replication and exit the cell will have a higher probability of infecting another cell, either in the same dish or in the dish of the next infection. These "rapid" viruses may be so because they enter a cell more rapidly than others, or because they uncoat more effectively, or because their RNAs are translated more efficiently, or they replicate RNA more rapidly, or mature new infectious particles more rapidly or, in short, because they perform better than their competitors in a rate-limiting step in the viral replication cycle (Chapter 2). Unfit mutant genomes are eliminated in this continuous competition, in a process that is called negative or purifying selection (see also Chapters 5 and 7). Mutations do not accumulate blindly in a replicating genome because many of the variant genomes that would accumulate debilitating mutations (or combinations of very slightly debilitating mutations) are never seen. They are either eliminated or kept at such low levels that they do not affect the average or consensus nucleotide sequence (compare this discussion with that on quasispecies in Chapter 4). It is worth noting once again that the reason why an RNA genomic region may be subjected to negative selection when mutations are introduced need not be related to its protein-coding potential. An RNA stretch, for example a foreign RNA inserted into an engineered clone, may not tolerate mutations because it plays a role in RNA secondary or tertiary structure or as a spacer in the newly generated replicon (Chapter 3).

Several types of mutations can be distinguished according to the nature of the modification in the nucleic acid, or the consequences for the encoded proteins, or their effect on viral fitness. They are summarized in Table 6.2. Some of the statements made in this Table are important to an understanding of RNA virus evolution, although they may seem obvious to molecular virologists. Most notably, a silent mutation need not be neutral, and neutrality can only be defined in a particular environmental context. This will become more apparent in the discussion on fitness variation in the next Chapter. Yet the assumption that silent mutations are selectively neutral has often strongly influenced population genetics during several decades.

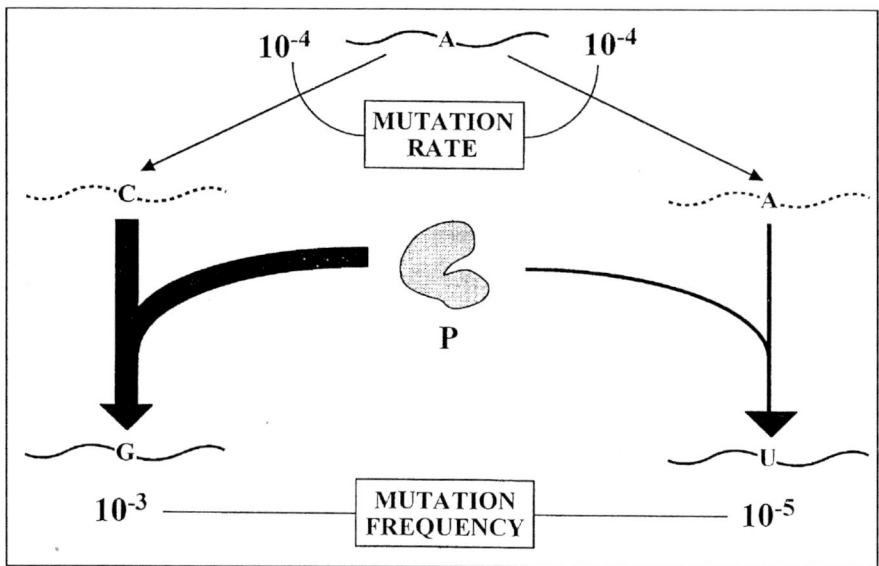

Fig. 6.3. Illustration of one example of a difference between mutation rate and mutation frequency. An A residue on a template is miscopied into C or A (instead of U) once every 10,000 times that the residue is copied. The complementary strand (discontinuous line) containing C is much more efficient that the strand containing A with regard to binding of a protein factor (P) required for further replication. The result will be a higher frequency of products with G than U. Mutation rates cannot be inferred from mutation frequencies (see text).

Fitness and neutrality are necessarily relative concepts, and values for the same biological entity often change in space and time.

A numerical example regarding mutation frequencies (or mutation rates if we assume that the particular mutations produced are perfectly accepted in progeny RNA) may clarify the biological impact of mutations for RNA viruses. A mutation rate of 10^{-6} substitutions per nucleotide copied (s/nt/rc) operating during replication of a 10 Kb genome would on average produce one mutation for each one hundred progeny RNA. If the average mutation rate were 10^{-4}, as in most estimates of mutation rates for RNA viruses, then one mutation would be present on average in each progeny genome. The difference is that to produce a selectively important mutation (for example an antibody-escape mutant which we assume needs only one mutation and the corresponding amino acid substitution) one hundred more progeny virus would be required in the former case than in the latter. This may not be an important difference in natural infections attaining high viral loads and rapid virion turnovers (in the range of 10^{10} to 10^{12} virions produced per day, as in HIV-1 or hepatitis B or C infections), but the difference may have great significance in some types of chronic or persistent infections involving limited numbers of infected cells, or limited permissivity of cells to the virus (Borna virus, for example). We may say that in these cases the exploration of sequence space by the virus, and consequently its evolutionary potential, and its long-term survival potential, are diminished. Therefore, the impact of a basal rate of mutation or recombination varies according to the rate of genome replication, turnover of virus and of susceptible cells, population size of the virus and of the potential target cells in the host organisms, migration of virus from one site of an organism into another, inflammatory and immune responses, and many other influences.

Table 6.2. Types of mutations in RNA evolution

TRANSITIONS are the mutations that result in the replacement of a purine by another purine (G→A or A→G) or of a pyrimidine by another pyrimidine (C → U or U → C).
TRANSVERSIONS are the mutations that result in the replacement of a purine by a pyrimidine (A→C, A→U, G→C, G→U) or of a pyrimidine by a purine (C→A, C→G, U→A, U→G).
INSERTIONS, DELETIONS (sometimes abbreviated as "indel" mutations) are modifications of the genetic material consisting in the addition (or deletion) of one or several nucleotides. In open-reading frames (ORFs) insertions or deletions that do not involve a number of nucleotides which is a multiple of three will result in frame-shift mutations.
SYNONYMOUS (or SILENT) mutations are those that do not give rise to an amino acid substitution, due to the degeneracy of the genetic code (depicted in Fig. 3.3, Chapter 3).
NEUTRAL mutations are those that do not affect fitness in a given environment. SILENT mutations need not be selectively neutral.
SELECTIVE mutations are those that affect fitness (positively or negatively) in a given environment. (It is often used to refer to positive effects on fitness).

Recombination and Reassortment as Promoters of Large Evolutionary Jumps

Genetic recombination has often served to describe any event leading to new combinations of genetic material regardless of the underlying mechanism. In virology, recombination has been distinguished from genome segment reassortment (Fig. 6.1). A recombinant RNA genome is one in which a covalently linked genome has been generated from two or more segments belonging to different parental genomes. In contrast, a reassortant RNA genome is one in which an entire genome segment (or several segments) of a multipartite genome has been replaced by the corresponding segment(s) of a different parental virus. In this process no covalent attachment of parts of distinguishable RNA genomic molecules occurs. Genome segment reassortment has a conceptual parallelism with chromosome sorting during sexual reproduction. Reassortment is restricted to viruses with segmented genomes (typically the influenza viruses, but also a number of multipartite bacterial, animal and plant viruses described in Chapter 2). The diploid retroviral genome lends itself to formation of viral particles with two nonidentical RNAs, for example if one of the two copies has undergone a mutational event. Unequal diploid genomes are in this case a step in the process of formation of recombinant genomes through strand transfers during reverse transcription. Recombination in RNA viruses is most frequently the result of a copy-choice (or template switching) during RNA or cDNA synthesis (Kirkegaard and Baltimore, 1986; Lai, 1995; Nagy and Simon, 1997). This implies that at least one of the parental coinfecting genomes must replicate in the cell for recombination to occur. Recently A. Chetverin, V. Agol and colleagues have obtained evidence that nonhomologous recombination between nonreplicating bacteriophage Qβ or poliovirus RNA genome regions can also occur. In the Qβ system, cell-free recombination was observed between the 5' and 3' fragments of a replicable RNA, probably guided by RNA secondary structure (Viebricher and Juce, 1998; Chetverin et al, 1997). Some of these recombination events have been attributed to ribozyme hammerhead-like structures that may form with low frequency in viral RNA molecules. However, the impact of these new recombination mechanisms in natural recombination of RNA viruses is not yet known.

Some viruses (alphaviruses, HIV-1, certain bacteriophages) engender a phenomenon known as superinfection exclusion. It is due to molecular interferences acting in such a way that when the cell is engaged in active replication of one virus, it cannot start efficient replication of a second, (usually somewhat related) superinfecting virus. This exclusion limits the possibilities of recombination that in these cases requires the simultaneous, synchronized coinfection with the two parents. For this, and also for other reasons, RNA viruses vary tremendously in their capability to yield recombinant genomes.

Homologous recombination occurs at high frequency in many positive strand RNA viruses both in cell culture and in vivo. Intertypic (among viruses of different serotype) and intratypic (among viruses of the same serotype) poliovirus recombinants are often rescued from vaccinees as a result of replication of attenuated vaccine strains in the gut (Minor, 1992). (Pathogenic viruses such as poliovirus are often divided into several serological types according to their cross-reactivity with the antibodies that they induce in host animals). Recombination is also quite active in retroviruses, and it has acted as a major evolutionary force in the expansion of immunodeficiency viruses (Sharp et al 1994). Some natural isolates of viruses such as subtype E HIV-1, or the alphavirus Western equine encephalitis virus (WEEV), appear to have been generated as a result of recombination events. Nucleotide sequence comparisons identified viruses related to Eastern equine encephalitis and to Sindbis virus as the likely ancestor genomes of WEEV (Strauss, 1993). Homologous (viable) recombination is very infrequent among the nonsegmented negative strand RNA viruses such as the rhabdovirus VSV. The fact that VSV has been highly successful in infecting mammalian hosts, and as an arbovirus, suggests that frequent recombination is not a general requirement to ensure adaptability and long-term survival of a virus group. The reasons why homologous recombination is rare in some viruses and common in others are not well understood, but they may relate to the processivity properties of the replicase (the ability to proceed end-to-end while copying one strand without falling off the template), the requirement of *cis*-acting versus *trans*-acting functions during replication, the microenvironment in which individual replicating complexes are located, and other influences. Because of the wide differences in recombination rates among different viruses, and the very variable recombination values for cellular DNA (due to different cell systems and tests used) found in the literature, it is not possible to compare recombination frequencies for RNA viruses and for DNA organisms, nor the relative impact that recombination may have in their evolution.

Recombination between parental genomic molecules with completely different nucleotide sequences is termed nonhomologous recombination. It has been described between some RNA virus genomes and cellular RNAs such as ribosomal RNA, tRNAs or messenger RNA. The chimaeric RNAs formed may be defective (Monroe and Schlesinger, 1983) or viable. Cytopathic forms of bovine viral diarrhea virus (BVDV) or virulent influenza virus have been described that had incorporated in their genomes foreign RNA stretches via nonhomologous recombination (Meyers et al, 1989; Khatchikian et al, 1989). The cytopathic forms of BVDV can be the result not only of recombination events but also of point mutations in a viral protein (NS2) resulting in the expression of a proteolytic activity (NS3) (Kümmerer et al, 1998). Nonhomologous recombination has probably occurred between viruses of a different family. One example is the acquisition by some coronaviruses of a messenger RNA segment identified also in influenza virus type C (Luytjes et al, 1988). This acquired segment provided this coronavirus group with an additional protein for attachment to cells. Most frequently, chimaeric viral genomes produced by nonhomologous recombination are either defective or attenuated in a variety of biological environments. Their isolation during natural infections appears to be facilitated by a singular gain of fitness in a very particular environment. Variations in fitness of RNA viruses can be large and rapid but also transient (Chapter 7). This means that viruses can explore and adapt to different biological and physical environments within short time periods. Consider the conflicting requirements imposed on the viral capsid or envelope to protect the viral genome outside the cell, and to introduce it efficiently into the cytoplasmic milieu (Chapter 2), let alone the variety of cell types, defensive responses, etc. confronting a virus during its multiplication in differentiated organisms. Too much adaptation to a particular set of environmental conditions may render the virus unable to multiply in other, related and frequently-ecountered environments. In terms of Wrightian fitness landscapes (Chapter 7) this specialized adaptation may be viewed as a virus being at the top of an isolated fitness peak in a constant environment. To reach other peaks the virus must descend to a valley and then search for new routes to high fitness peaks. The best situation for a virus is to find itself at a peak belonging to an extensive mountain range with many peaks accessible without the need to descend to low valleys.

The contribution of RNA recombination to the evolution of RNA viruses has been interpreted in two opposite ways. In one scenario it has been regarded as a means to promote large evolutionary jumps among RNA genomes. Many nonhomologous recombination events must be unsuccessful evolutionary trials. A few, however, such as the case of WEEV discussed in preceding paragraphs, may lead to the emergence of a new RNA pathogen. In fact, a good proportion of the emerging and reemerging viral diseases (Chapter 8) are associated with RNA viruses showing active recombination. A second, differing interpretation, is that recombination allows for the rescue of viable, fit genomes from defective or highly debilitated parents. In both interpretations, successful RNA recombination may be regarded as promoting an evolutionary jump which takes a virus to a new fitness peak.

One important question which applies to mutation, recombination and segment reassortment is whether they occur at random or rather preferentially at some genome sites, or if they are subjected to some type of regulation. Evidence for mutation is that it is not entirely random since some mutational "hot-spots" (Table 6.1) have been documented both in vivo and in vitro. However, selection among the many primary mutants generated during genome replication appears to be the major process that determines the repertoire of variants that we can identify in viral populations. Probably, similar arguments apply to recombination, although the degree of nonrandomness may vary from one viral system to another (compare Banner and Lai, 1991; Rowe et al, 1997; Anderson et al, 1998). Finally, this question has not been settled for genome segment reassortment, and the two extreme models are that grouping of genomes is carefully regulated to ensure the required genetic complement, or that it occurs at random and only the valid combinations can initiate a productive infection.

Rate of Accumulation (or Fixation) of Mutations, and Its Lack of Regularity

The rate of genome evolution in nature, that is, how the genetic composition of a virus population changes with time, can be measured to a first approximation as the rate of accumulation of mutations in the viral genome consensus sequence with time (Table 6.3.). Determination of the rate of evolution involves comparing consensus genomic sequences of sequential isolates (for example hepatitis C virus from successive isolations of a chronically infected patient) or of independent isolates belonging to the same evolutionary lineage (for example isolates of the same virus from several infected hosts from one geographical area at year-long intervals). Needless to say that, as documented several years ago with the important pathogen foot-and-mouth disease virus (FMDV) (Sobrino et al 1986), rates of fixation of mutations can exhibit an impressively large range of values even within a single disease episode. They vary depending on the particular viral gene or gene segment considered, and also on the particular time span of the observations. Many viruses display periods of rapid evolution with intervening periods of genetic stasis ("punctuated equilibrium"). In FMDV rates as low as $< 4 \times 10^{-4}$, and as high as $> 10^{-2}$ substitutions per nucleotide position per year, have been measured depending on the genomic segments and time periods considered (Sobrino et al, 1986). Moreover, for the capsid protein VP1-coding region of FMDV, estimated rates of evolution were 1.9×10^{-3} substitutions per nucleotide position per year, and 4.6×10^{-2} substitutions per nucleotide

Table 6.3. Rates of evolution for RNA viruses

VIRUS	GENOMIC REGION	SUBSTITUTIONS PER NUCLEOTIDE PER YEAR	REFERENCE
Foot-and-mouth disease virus	Average over genome	$<4\times10^{-4}$-4×10^{-2}	Sobrino et al 1986; Domingo et al 1990
	VP1[a]	9.0×10^{-3}-7.4×10^{-2}	Gebauer et al 1988
	VP1	6.5×10^{-3}	Villaverde et al 1991
	VP1	1.4×10^{-4}	Martinez et al 1991
Poliovirus	Average over genome	6.9×10^{-3}-1.4×10^{-2}	Kew et al 1990
	VP1[b]	3×10^{-2}	Kew et al 1998
Enterovirus 70	VP1	5×10^{-3}	Takeda et al 1994
Eastern equine encephalomyelitis virus	26S RNA	1.4×10^{-2}	Weaver et al 1991
Hepatitis C virus	Average over genome[c]	1.4×10^{-3}	Okamoto et al 1992
		2×10^{-3}	Smith et al 1997
	Structural proteins[b]	9×10^{-4}	Abe et al 1992
	5'NC, C, E, NS1,2,3,5[d]	2×10^{-3}	Ogata et al 1991
Hepatitis G virus	5'NC[d,e]	4×10^{-4}	Gimenez-Barcons et al, 1998
	NS3[d,e]	2×10^{-3}	
	Average over genome	3.9×10^{-4}	Nakao et al 1997
Hepatitis delta virus	Delta antigen[d]	6×10^{-4}	Imazequi et al 1990
	Average over genome	3.0×10^{-2}–$3/0\times10^{-3}$	Lee et al 1992
Hemorrhagic septicemia virus (fish rhabdovirus)	G	1.6×10^{-4} (slow phase) 4.5×10^{-3} (fast phase)	Benmansour et al 1997
Influenza virus, Type A	NS	1.9×10^{-3}	Smith & Palese 1988
	HA(H3)	6.7×10^{-3}	Smith & Palese 1988
	HA(H3)	5.7×10^{-3}	Fitch et al, 1997
	HA(H1)[f]	5.8×10^{-3}-1.7×10^{-2}	Rocha et al 1991
	NA(N2)	3.2×10^{-3}	Smith & Palese 1988
	M1	8×10^{-4}	Ito et al 91
	M2	1.4×10^{-3}	Ito et al 91
	NP	1.6×10^{-3}-2.2×10^{-3}	Altmuller et al 1989 Gorman et al 1990
	NP[e]	3.5×10^{-3}-2.4×10^{-2}	Rocha et al 1991
Influenza virus, Type B	HA	3.4×10^{-3}-4.7×10^{-3}	Rota et al, 1990
	HA, NS	1.1×10^{-3}	Yamashita et al 1988
Visna virus	Average over genome	1.7×10^{-3}	Braun et al 1987
Visna-related	Average over genome (nonsynonymous subst.)	$8¥10^{-4}$	Querat et al 1990

Table 6.3. continued

VIRUS	GENOMIC REGION	SUBSTITUTIONS PER NUCLEOTIDE PER YEAR	REFERENCE
Human immunodeficiency virus	env	3.2×10^{-3}-1.6×10^{-2}	Hahn et al 1986
	gag	3.7×10^{-4}-1.8×10^{-3}	
Simian immunodeficiency virus	gp120[g]	8.5×10^{-3}	Burns and Desrosiers 1991
Equine infectious anemia virus	Env[h]	10^{-1}-10^{-2}	Clements et al 1988
HTLV-II	LTR[i]	1×10^{-4}-3×10^{-5}	Salemi et al 1998
Canine parvovirus	VP1-VP2	1.7×10^{-4}	Parrish et al 1991
Polyomavirus JC	Intergenic region	1×10^{-7}-3×10^{-7}	Sugimoto et al 1997
Papillomavirus	E6	3×10^{-8}	Van Ranst et al 1995
Cellular genes of hosts		10^{-8}-10^{-9}	Britten, 1986; Weissmann and Weber, 1986

[a] Persistent infection established experimentally *in vivo*
[b] Synonymous mutations; immunodeficient patient
[c] Human isolate administered to chimpanzee
[d] Intrapatient evolution
[e] Average of many patients coinfected with HIV-1
[f] Immunodeficieny child persistently infected with IV
[g] Infections with a molecular clone of SIV
[h] Chronically infected ponies
[i] Drug users

position per year, when measured over a nine or a two year interval, respectively. This broad range of values spanning two orders of magnitude was calculated by comparing viral isolates from acutely infected animals. In experimental, persistent infections of cattle established with a biologically cloned FMDV, the values within individual animals ranged from 6.9×10^{-3} to 1.4×10^{-2} substitutions per nucleotide per year (Gebauer et al, 1988). Likewise, in a number of studies with human influenza virus type A, rates of fixation of mutations ranging from 8×10^{-4} to 2.4×10^{-2} substitutions per nucleotide per year have been measured for different hosts and different biological environments (Table 6.3.). Furthermore, for the hemagglutinin gene of influenza B virus, the rate of fixation of mutations was calculated to be 4×10^{-3} substitutions per nucleotide position per year with viruses isolated over a ten year period (Rota et al, 1990) and 1×10^{-3} substitutions per nucleotide position per year over a 47 year period (Yamashita et al, 1998). Low rates have been estimated for influenza A viruses that colonize natural bird hosts without causing pathology (Webster et al, 1995). This provides an example of evolutionary stasis in nature that contrasts with the rapid evolution of the same types of influenza viruses when they infect humans (Table 6.3). Birds and swine are important reservoirs of influenza viruses with the potential danger that they may become human pathogens, either as they are found in these animals, or after genome segment reassortment with circulating human viruses. A phylogenetic analysis of the surface glycoprotein of the fish rhabdovirus viral hemorrhagic septicemia virus revealed two phases of evolution (Benmansour et al, 1997). A rapid evolutionary phase with rates of fixation of mutations of 4.5×10^{-3} substitutions per nucleotide position per year, and a slow phase (genetic stasis) with rates of 1.6×10^{-4} substitutions per nucleotide position per year. It must be pointed out that in RNA virus evolution, genetic stasis (as observed in the case of this fish rhabdovirus or with some arboviruses) is often represented by rates of evolution in the range of 10^{-4} substitutions per nucleotide position per year, still many orders of magnitude higher than the rates that have been estimated for cellular genes (Table 6.3).

In a patient chronically infected with HCV, the average rate of fixation of mutations was estimated to be 2×10^{-3} substitutions per nucleotide position per year (2.5% nucleotide differences in a 13 year period), but in a 39-nucleotide stretch of the NS1-coding region the difference amounted to 28% of nucleotides (Ogata et al, 1991). A striking absolute conservation in the 3' NC and 3D (polymerase) gene versus mutation frequencies of about 10^{-3} substitutions per nucleotide position elsewhere in the genome have been detected during persistent infections of FMDV in cell culture (Toja et al, 1999).

From the data summarized in preceding paragraphs, and also supported by a variety of studies compiled in Domingo and Holland, 1994; Meyerhans and Vartanien, 1999; Suzuki et al, 2000, and in Table 6.3, it must be concluded that rates of evolution for RNA viruses are generally rapid but quite variable and irregular, and they cannot serve accurately to infer times of divergence of related viruses, nor the time at which a particular type or strain of virus originated. When there appears to be some kind of "molecular clock" (a steady accumulation of mutations as a function of time) it is probably often fortuitous. Rapid rates of accumulation of mutations may often have very limited meaning except for short, defined time periods in which they may reflect divergence from a focal origin or a directed selective pressure applied to a replicating viral population. For example, they may indicate rapid variation of an antigenic site as a reflection of positive immune selection, or rapid variation of a viral enzyme as a result of an antiviral intervention with inhibitors directed to that particular enzyme. Differing applications of phylogenetic techniques to date the origin of viruses has created (and is still creating) considerable confusion. Let us just mention, for example, the proposal that HIV-1 may have diverged from HIV-2 only 50 years ago (Smith et al, 1988), or 200 years ago (Querat et al, 1990), while application of recently developed procedures based on statistical geometry gives about 1000 years for the same time point of divergence (Eigen and Nieselt-Struve, 1990). RNA genetics differs in a number of respects from classical population genetics as applied to differentiated organisms, and the lack of a dependably steady accumulation of mutations is one of them. This does not mean that there is agreement on the existence of a dependably regular molecular evolutionary clock for differentiated organisms. Since this concept was first proposed by Zuckerkandl and Pauling (1965) evidence against it has abounded in the literature. However, it must be stressed that for RNA viruses, the very nature of their replication mode at the population level, renders very unlikely the existence of a regular molecular clock regarding consensus sequences, at least for extended time periods.

A number of comparisons of animal and plant genes and gene families suggest rates of evolution in the range of 10^{-8} to 10^{-9} substitutions per nucleotide position per year. The rapid evolution of RNA genomes, as compared with DNA genetic elements was remarkably illustrated by the million-fold greater rate of accumulation of mutations of the *v-mos* gene of Moloney murine sarcoma virus (1.3×10^{-3} substitutions per nucleotide position per year) than its cellular counterpart, c-mos (Gojobori and Yokoyama, 1985). Another example is provided by HTLV-1 and HTLV-II, two relatively static human retroviruses. In HTLV-1 the available evidence suggest viral amplification by clonal expansion of T cells bearing integrated provirus, with limited virus particle formation (Wattel et al, 1995; Wain-Hobson, 1996). This virus has attained a maximum genetic divergence of merely 7% of its genomic nucleotides, and viral sequences have served to trace human migrations (Gessain et al, 1995). For HTLV-II, rates of virus evolution among drug users appear to be about one order of magnitude slower than the average values commonly estimated for RNA viruses (Salemi et al, 1998). A number of evolutionary rates for DNA viruses are included in Table 6.3. They are closer to the rates estimated for cellular genes than for RNA viruses. It has been suggested that some of these viruses, such as the polyomavirus JC, are more adequate to trace human migrations than mitochondrial human DNA, since variability of the latter may obscure the tracing of human lineages (Sugimoto et al, 1997).

Population Equilibrium: Stasis in the Face of High Mutation Rates

It is generally accepted in genetics that the probability of any mutation being detrimental is larger for organisms well adapted to their environment than for organisms that are poorly adapted. In the case of viruses, let us assume that a viral genome has been replicating in the same environment for a large number of generations. This was to some extent what happened to bacteriophage Qβ since its isolation in Kyoto in 1961. At that time scientists were not aware of the likely genetic changes that a virus may undergo upon passage, and it was common to routinely passage viruses or bacteria to produce new stocks as needed. Bacteriophage Qβ was passaged many times in the laboratories of Watanabe (Tokyo), Spiegelman (Urbana) and Weissmann (New York and Zürich). In all cases the host for Qβ was *E. coli*, and although there might have been minor environmental variations (of *E. coli* strain, composition of the culture medium, or even in temperature due to faulty calibration of thermometers...) it is fair to assume that the environment was reasonably invariant. The successive passages of bacteriophage Qβ conformed to an unintended long-term evolution experiment. A sampling of nucleotide sequences of clonal Qβ preparations as well as of multiply passaged populations indicated that phage populations were genetically heterogeneous and consisted of a mutant spectrum with individual component genomes carrying an average of one to two mutations relative to the consensus. In other words, the invariant Qβ genomes (or the zero mutation class) amounted to only about 15% of the population, the remaining genomes deviating in one or more mutations from the zero mutation class (Domingo et al, 1978). Yet, in spite of this remarkable heterogeneity which in fact provided the first experimental indication of the quasispecies structure of an RNA virus, the average or consensus nucleotide sequence of the Qβ genome remained invariant for many generations, in a number of controlled, serial passage experiments. This means that despite mutant genomes arising at a high rate [as documented by the rich mutant spectrum and by a measurements of mutation rate for a specific base transition (Batschelet et al, 1976)], none of the individual variants increased its amount sufficiently to modify the average. This situation of invariance of a consensus or average genomic nucleotide sequence in spite of continuing replication and mutant generation is termed population equilibrium. At equilibrium, an RNA virus exists as a weighted average of many related sequences that constitute the mutant spectrum of the viral quasispecies (this is in conformity with the mathematical formulation in Chapter 4).

The average or consensus nucleotide sequence of an RNA genome population represents the result of rating the competitive fitness of a large number of mutants that arise and compete as they complete infectious cycles. When a virus is well adapted to a biological and physical environment, replication will produce variant genomes but each specific variant will remain at low frequency because the entire ensemble of mutants (the unit of selection) is nearly optimal under those circumstances. Negative selection is a factor in genome stability. It is the force that maintains a dynamic mutant spectrum leading repeatedly to the same average genomic sequence. Therefore an invariant (or slowly evolving) consensus sequence in vivo or in cell culture does not mean that a mutant spectrum and a dynamic quasispecies are absent. Examples of evolutionary stasis, defined by an invariant or slow-evolving replicating virus, will generally be due to population equilibrium rather than to the absence of a high mutation rate. The alternative possibilities are that the genome pool is either not replicating (or replicating very slowly) or its replication is mediated by a high-fidelity replication or repair machinery, a situation that has, as yet, never been documented for RNA viruses.

Table 6.4. Some factors promoting population equilibrium and disequilibrium in RNA virus populations

EQUILIBRIUM
- Constant environment
- Long-term adaptation to an environment
- Limited replication
- Large and rather constant viral population size

DISEQUILIBRIUM
- Within host organisms:
  - Inflammation, fever response
  - Immune selection (antibodies, CTLs, effectors)
  - Altered nutritional status
  - Immune debilitation
- External to organisms:
  - Changes in temperature and weather
  - Administration of inhibitors, vaccines (or other effectors that produce metabolic alterations)
  - Ecological and environmental modifications that alter host and viral traffic[a]
  - Host demographics
- Changes in viral population size
  - Random sampling or bottleneck events (within hosts, upon transmission)[b]
  - Size of mutant spectrum (variant repertoire)

[a] These influences are discussed with greater detail in Chapter 8 in connection with emergence and reemergence of viral diseases.
[b] See also Chapter 7.
Based in Domingo (1996, 1998) and Domingo and Holland (1988, 1992).

Population Disequilibrium: the Trigger of Rapid RNA Genome Evolution

RNA virus evolution, defined as a change in the average nucleotide sequence of the viral genome, occurs only when viral populations deviate from a particular population equilibrium. Factors that contribute to an evolutionary disequilibrium are (i) changes in environment, (ii) random sampling events and (iii) the stochastic emergence of rare genetic changes with an evolutionary advantage, be it by mutation or recombination (Table 6.4).

Environmental changes often perturb population equilibrium. As an example, the introduction of neutralizing antibodies directed to one specific antigenic site of a replicating virus will confer a selective advantage to those components of the mutant spectrum with amino acid replacements that render the virus less sensitive to the neutralization action of the antibodies. This is an example of population disequilibrium which has been extensively documented during viral infections in vivo and in cell culture. The perturbation of population equilibrium is manifested by a modification of the consensus genomic sequence which will now be dominated by variants with diminished sensitivity to the antibodies. Accommodation to a new environment need not be a single-step process. First a single mutation that renders the virus partially resistant to the antibody molecules may dominate after very few generations, either because the critical mutation preexisted in some minority components of the mutant spectrum or because the mutation may arise readily during replication. In this population enriched with the single mutant, a second mutation may take place that enhances the selective advantage that causes a further shift of the mutant distribution. The same two-error mutant could have been produced in a single replication event, although at a much lower rate: if the probability to find a specific one-error mutant is 10^{-4}, a two-error mutant is found with a probability of 2×10^{-8}, a three-error mutant with a probability of 4×10^{-12} and so on (Chapter 4), provided that the mutation events are independent one from another. All steps leading uphill in the fitness landscape are allowed, but one or even two level steps (or even slightly downhill) are tolerated. Adaptation to an environment is never complete (as with human happiness or world peace), it eventually ends at the nearest-by local optimum where a new evolutionary equilibrium is established, until this equilibrium is again perturbed. A variety of environmental influences have been documented to promote population disequilibrium in infections in vivo and in cell culture (Table 6.4).

It is worth noting that the difference between positive and negative selection becomes blurred as we picture the events leading to an adjustment of the quasispecies following an environmental change. The dominance of antibody-resistant mutant distributions can be viewed in two different ways: the positive selection of the fitter mutants or the negative selection of the invariant ones. Dominance and inferiority are relative, as can be observed for multi-team matches in sporting events.

A second mechanism of perturbation of population equilibrium is through random sampling events in which only one or a few individuals of a mutant distribution are the ones allowed to reproduce to form a new distribution. This probably occurs frequently when one or a few viral particles are transmitted from an infected individual to a susceptible one, via respiratory droplets or other small volumes of infected material. This founder event may be a source of disequilibrium because even if the environment of the recipient organism were identical to the environment of the donor (an unlikely assumption) the quasispecies must be rebuilt from a few founder genomes. If the latter deviated from the previous optimum, it may take several rounds of replication (walk in sequence space) before a new population equilibrium is reached. Differences in the fine genetic make-up of the infecting virus, together with physiological and immunological parameters of the host, are both likely to contribute to the different severity of viral infections in different individuals.

Population equilibrium cannot be regarded as a steady, permanent situation. This statement is based on the continuous exploration of sequence space by replicating viruses, and the essentially stochastic nature of mutagenesis as well as the instability of environments. Let us assume that a virus is replicating in a constant environment and that an equilibrium has been reached as judged from the invariance of the consensus genomic nucleotide sequence. This equilibrium has been attained after continuing exploration of sequence space by testing mutant distributions arising from mutation. This sequence exploration is limited for statistical reasons. Single, double, triple mutants (or perhaps higher order mutants if the replicating population size is sufficiently large) can be easily explored upon replication. However, there are mutants (for example, one that differs precisely at 50 specific sites from the average genome) that do not appear (negligible probability) during replication in one or a number of infected hosts. Such unlikely precise combinations of mutations might occur say once every 10^6 years of replication, far more than the existence of the host or the virus itself. Yet this potential 50-position mutant may represent a better optimum than the current distribution. Many 50-position mutants of HIV-1 are explored in infected patients! If instead of 50 specific mutations, only 10 to 20 are needed for a new optimum the occurrence of these rare genomes may be sufficiently frequent as to be observed as perturbations of equilibrium in spite of a constant environment. Only with the theoretical assumption of an infinite population or an infinite number of generations of passage history (in which case all possible genomes could be tested) could a true, stable population equilibrium be guaranteed (compare with parallel conclusions with simple RNA replicons in Chapter 5). Thus, viral quasispecies are finite mutant distributions subjected to perturbations of equilibrium due to selective events and stochastic effects (Eigen and Biebricher, 1988; Domingo et al, 1985, 1995).

Unequal Rate of Occurrence of Different Mutation Types

It is important not to extrapolate to the entire genome mutation rates based on determinations for a specific nucleotide site. As discussed in Chapter 3, a number of influences affect misincorporation rates at specific genome sites. As an example, Sedivy et al (1987), in an elegant study of poliovirus genetics, calculated a mutation rate of 2×10^{-6} substitutions per nucleotide for the reversion of an amber mutant that required a specific transversion within the polymerase open-reading-frame of poliovirus RNA. They emphasized the unusual low value estimated by their genetic method and suggested a rather low mutation rate for poliovirus. However, it was later shown that during replication the poliovirus polymerase introduces transition mutations with a 50-fold higher frequency than transversion mutations (Kuge et al, 1989). This correction brings the estimated average mutation rate for poliovirus within the range of values determined for other poliovirus mutations and for RNA viruses using a variety of genetic and biochemical procedures [reviewed in Drake, 1993; Domingo and Holland, 1994; Drake et al, 1998; Drake and Holland, 1999].

A word of caution is needed before interpreting mutation rates based on repetitive nucleotide sequencing of genomes produced during the development of a plaque (the progeny from a single genome on a cell monolayer) [values obtained by this procedure were reviewed by Domingo and Holland (1994)]. Any mutation arising during plaque development which is not at least neutral will be overgrown by those dominant genomes which will be expanding during the growth of the plaque. Even if a mutation were strictly neutral it must occur quite early during plaque development to be scored as a mutation in the sequencing of the population of genomes sampled from the plaque, or to have a chance to be picked for the next plating or amplification prior to sequencing (the situation is parallel to that depicted in Fig. 4.3 in Chapter 4 to illustrate the classical mutant generation experiment of Luria and Delbrück). In more general terms, since the majority of newly arising mutants will tend to show lower fitness than their fit (highly selected) "wild type" parental genomes (Chapter 7), most determinations of mutant frequencies—in which the relative fitness of the mutants intervenes in the final populations being analyzed—are likely to be underestimates of true mutation rates [this important point was already emphasized by Domingo and Holland (1994)]. Obviously, neutral mutants may also lead to overestimates of mutation rates after many replication rounds due to error propagation and fluctuations in the mutant level due to random drift.

Transitions are generally more frequent than transversions during short-term evolution of RNA viruses, such as in the expansion of populations from a clonal origin. In these cases, point mutations tend also to be more frequent than insertions or deletions, probably because of their increased rate of occurrence as well as for their lower probability of adversely affecting viral fitness. Again, it must be emphasized that the rate of occurrence of specific types of mutations must be distinguished from the frequency with which they are found in a viral population. Some types of mutations may only be observed when certain passage conditions of the virus are fulfilled. A specific example, discussed in more detail in Chapter 7, is an unusual internal polyadenylate tract found in foot-and-mouth disease virus (FMDV) populations derived from serial plaque transfers. Such a tract had never been observed previously in other FMDVs, either natural isolates or cell culture adapted populations. Yet, the present evidence qualifies the site of the tract as a mutational hot spot in the FMDV genome—a hot spot which regularly debilitates progeny so that this internal polyadenylate is never observed in FMDV multiplying under usual laboratory conditions or in animal hosts.

Hypomutation and Hypermutation

Mutation rates of 10^{-3} to 10^{-5} substitutions per nucleotide copied are average values that do not necessarily reflect the rates at individual sites of RNA viruses. Because of numerous effects (Chapter 3), mutation rates at individual sites may be either lower (hypomutation) or higher (hypermutation) than the average.

Striking cases of hypermutated RNA molecules were identified in defective-interfering (DI) RNAs of VSV (O'Hara et al, 1984b), and in variant measles viruses persisting in the central nervous systems of humans, associated with the lethal syndromes measles inclusion body encephalitis (MIBE) and subacute sclerosing panencephalitis (SSPE). In the brain of these patients no infectious measles virus is detected. However, defective viruses can be isolated which contain multiple genetic lesions and defects in gene expression (review in Cattaneo and Billeter, 1992). Increased frequencies of U → C or A → G transitions were detected in some genomic regions of the measles variants persisting in the brain. This phenomenon is known as biased hypermutation. It is thought that such mutation events might contribute to long-term persistence of deviant measles virus genomes by aborting particle formation and minimizing exposure of infected cells to immune attack. Hypermutation could either contribute to persistence or be the result of release of functional constraints in those regions of the variant measles genomes which are concerned mainly with maturation and infectivity, but not with genome replication.

Several models have been proposed to explain biased hypermutation. One is the active enzymatic editing of the viral RNA e.g., in the form of a double stranded RNA unwindase that is able to deaminate and oxidize up to 50% of adenosine residues in double stranded RNA to produce inosine (the ribonucleoside form of hypoxanthine). Since inosine directs the incorporation of cytidine phosphate, further viral replication would result in the massive biased mutation events that are observed (Cattaneo and Billeter, 1992). For human immunodeficiency virus, dislocation mutagenesis (that is, slippage of the primer relative to the template during reverse transcription) has been proposed to account for G → A biased hypermutation (Vartanian et al, 1991). A bias in favor of G → A substitutions was also observed in the mutant spectrum of satellite RNAs of the plant virus tobacco mosaic virus (Kurath et al, 1992).

Hypermutation has been identified in viruses with disparate replication strategies such as human parainfluenza and respiratory syncytial virus, retroviruses and hepatitis B virus. Although the phenomenon often affects defective genomes, in the case of respiratory syncytial virus, viable hypermutated genomes have been described for infectious variants that are able to escape neutralizing antibodies (Rueda et al, 1994).

A number of in vitro hypermutagenesis procedures have been developed to allow insightful and potentially useful exploration of the functional space of nucleic acids and proteins (Leung et al, 1989; Martínez et al, 1996; Vartanian et al, 1996; Zaccolo and Gherardi, 1999; see also Chapter 8).

Genetic Heterogeneity of Natural Populations of RNA Viruses

The term genetic heterogeneity of RNA virus populations has been used widely to describe two quite different sets of values. One is the genetic distance (or Hamming distance, defined in Chapter 4) when comparing independent isolates of the same virus from different infected hosts (or geographic regions). The other meaning is the extent of variation (genetic distance) among the individual genomes that compose a given viral isolate or clone. A general observation has been that independent isolates of the same virus serotype are often genetically unique. This was suggested by results of an early study comparing natural isolates of VSV (Clewley et al, 1977), and in the first series of analyses carried out in Madrid using human influenza virus and foot-and-mouth disease virus (FMDV) (Ortín et al, 1980; Domingo et al,

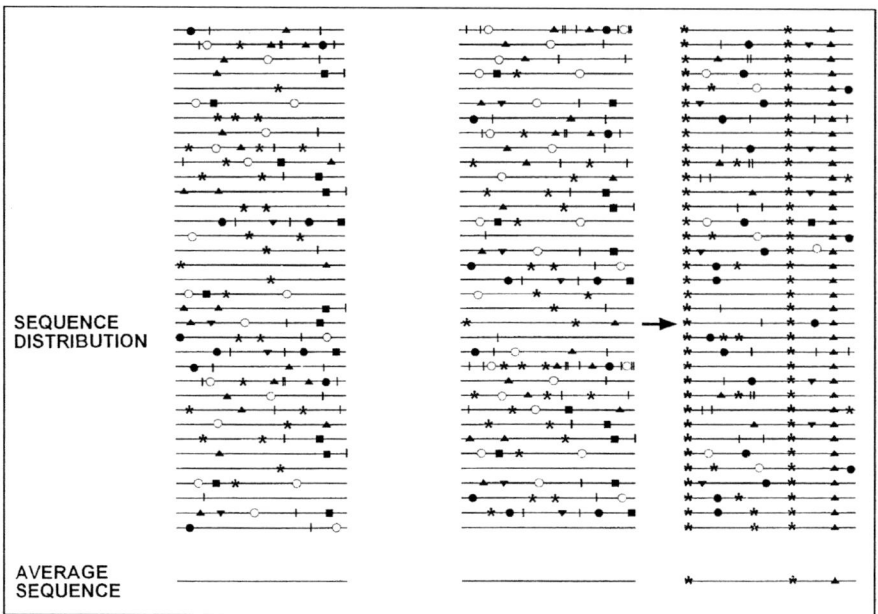

Fig. 6.4. Schematic representation of a viral quasispecies. Genomes are represented by horizontal lines, and mutations as symbols on the lines. RNA virus populations consist of dynamic spectra of mutants. The arrow represents the amplification of a single genome from the middle distribution to yield a new quasispecies with a new average sequence. Note that, depending of the complexity of the distribution (average number of mutations per genome) the average or consensus genome may not have a physical reality in the distribution (see text).

1980). In this early study with FMDV, the extent of genetic heterogeneity among independent viral isolates from Spain was estimated to be 0.7-2.2% of the genomic nucleotides. Heterogeneities ranging from 1-20% nucleotide divergence have been estimated in several studies in which sequences of independent FMDV isolates from a given geographical area were compared. In the early analyses of FMDV isolates, heterogeneity was also detected among viral genomes within an infected animal, which led to the proposal of a quasispecies structure for FMDV in vivo (Domingo et al, 1980). The extent of heterogeneity within this isolate was 1% of the heterogeneity observed among the consensus genomic sequence of independent isolates. This intraisolate heterogeneity represented an average of about one mutation per infectious genome. More recent results suggest that the FMDV genome heterogeneity in an infected animal was probably underestimated in these early studies (Carrillo et al, 1990).

Examination of individual biological or molecular clones from viruses as they infect their natural hosts systematically reveals mutations that distinguish the individual genomes which compose the viral population (Fig. 6.4). The degree of heterogeneity within each isolate is variable in magnitude. Studies have involved a great number of human, animal and plant viruses. In recent years, direct nucleotide sequencing after cDNA copying and polymerase chain reaction amplification (a procedure termed RT-PCR, see Chapter 8) of one or several genomic regions has become a standard technique. However, several sampling procedures (T1-oligonucleotide fingerpringint, RNase A mismatch detection, restriction fragment length polymorphism, heteroduplex mobility assays, etc.) have contributed, and are still contributing, to document the extensive heterogeneity among independent isolates of the same RNA virus or among individual genomes of the same isolate. The reader can find numerous examples of these types of analyses in most issues of virology journals, and additional examples were reviewed in books covering topics of viral evolution (Domingo et al, 1988, 1999; Holland, 1992; Morse, 1994; Gibbs et al, 1995).

Once different sequences are available either from one or several infected individuals, they can be analyzed using classical phylogenetic procedures [maximum parsimony, maximum likelihood, neighbor joining, split decomposition or other (Huelsenbeck and Rannala, 1997; Page and Holmes, 1998), available in several computer program packages]. Phylogenetic analyses have contributed to defining the mode and tempo of viral evolution within infected organisms and to the establishment of relationships among isolates. This allows a definition of types, subtypes and clades on the basis of genotypic differences. For example, subtypes of HIV-1 have been based on *env* gene sequences using phylogenetic procedures, and HIV-1 subtypes have been shown to be distributed in a nonuniform way in different geographic locations: subtype B in Europe and the U.S., D in Africa, C and E in Asia. However, because of human mobility, the geographic association of HIV-1 subtypes is changing with time. Phylogenetic comparisons are also useful in molecular epidemiology, to trace the origin of new virus outbreaks. The use of nucleotide sequencing and phylogenetic methods to study RNA viruses has documented very high diversity among independent isolates, as well as a continuous process of genetic diversification within infected hosts.

A few recent examples will illustrate that even viruses traditionally regarded as genetically stable and serologically monotypic participate in a quasispecies structure, and that genetic heterogeneity may have dramatic consequences for the biological behavior of a virus. Rabies virus has been considered antigenically monotypic, and a classical vaccine derived from the time of Louis Pasteur has been used successfully for over a century. Many workers have long considered genetic variation of rabies virus to be of little relevance to its biology. Yet A. Flamand and colleagues (Benmansour et al, 1992) provided the first evidence of the quasispecies nature of rabies virus. Sequence comparisons of rabies viruses world-wide have distinguished seven genotypes, each occupying a rather defined geographical location. The classical vaccine is efficient for protection against rabies caused by some genotypes but not by others. New, more complex, multivalent vaccines are currently under investigation. Furthermore, the presence of two genetically and biologically distinct variant rabies viruses has been documented by passage of a standard challenge strain in BHK cells (Morimoto et al, 1998). The surface glycoprotein of the two variants differs in ten amino acids, and they exhibit distinct cell tropisms and pathogenic potential for mice. The variant that dominates the challenge virus standard is more neurotropic in vivo and in cell culture than the variant selected in BHK-21 cells. It is also more pathogenic for adult mice, whereas the BHK-selected variant is more pathogenic for neonatal mice. It is now increasingly likely that variants hidden with the mutant spectra of rabies virus quasispecies may contribute to changes in host cell tropism, and to the emergence of new forms of this pathogen. This could be true for silver-haired bat rabies virus, associated with human rabies cases in North America in the 1990's (Rupprecht et al, 1995). Also variants differing in neuropathogenicity for mice have been isolated from lactate-dehydrogenase-elevating virus quasispecies (Chen et al, 1998).

Another example concerns current research on cell receptors for measles virus (MV). Many studies with MV, including identification of the surface antigen CD46 as the MV receptor, have been conducted with a vaccine strain of MV called Edmonston. This attenuated strain had been passaged extensively in monkey kidney cells. Surprisingly, marmosets lacking the essential CD46 virus-binding motif were nevertheless susceptible to a number of natural isolates of MV but not to the Edmonston vaccine strain (Hsu et al, 1998, and references therein). It has been shown that a single amino acid replacement (Asn-481 → Tyr) in the surface hemagglutinin of the viral envelope allows MV to bind CD46. Tyr-481 is present in the Edmonston strain, and Asn-481 is present in most wild type isolates. A second receptor, other than CD46 serves for the binding of MV to B-lymphocytes. It was only due to the occurrence of an amino acid replacement, presumably during cell culture adaptation of the Edmonston strain, that CD46 was discovered as a MV receptor, although CD46 may not be the authentic major receptor for many natural MV isolates. It is paradoxical that such a striking change in receptor specificity, dependent on a single amino acid replacement, has been unveiled for a virus generally regarded as genetically and antigenically stable, and for which a classical vaccine has been successful during several decades. Yet, even for this seemingly invariant agent, a single amino acid change was sufficient to create a problem regarding identification of its cellular receptor.

The recently discovered flavivirus-like hepatitis G virus shows a quasispecies structure, but it is overall more conserved than hepatitis C virus (Pickering et al, 1997; Viazov et al, 1997). Yet the intrapatient rate of evolution of hepatitis G may reach values of 2×10^{-3} to 4×10^{-4} substitutions per nucleotide position per year (Nakao et al, 1997; Gimenez-Barcons et al, 1998), which are typical of RNA viruses (Table 6.3).

The analyses of both biological clones and molecular clones of RNA viruses converge to indicate that strictly homogeneous RNA viral populations must be very rare. These same results also indicate, however, that the extent of genetic heterogeneity may vary up to two orders of magnitude depending on the viral system and the type and history of the infection. Particularly important is the number of replication rounds undergone by the virus since its last bottleneck event (amplification from one or few particles). Average values of one to ten mutations per genome were estimated by F. Sobrino and his colleagues on replication of FMDV in swine (Carrillo et al, 1990). Values approaching one hundred mutations per genome have been estimated as the differences among individual human immunodeficiency type 1 genomes present in infected patients after prolonged infection (Nájera et al, 1995; Coffin, 1995).

Genetic heterogeneity per se, without prior knowledge of the history of the infection, is not sufficient evidence for a quasispecies behavior. In some early reports, the identification of closely related viral genomic sequences in an infected organism was interpreted as being the result of coinfection with two different viruses, reflecting how reluctant virologists were to accept mutation as a frequent event during viral replication. When the mutant composition of a viral population (particularly one generated from a clone) is followed as a function of time, the dynamics of mutant generation, competition and selection, the hallmark of quasispecies behavior, has been demonstrated.

Quasispecies Dynamics in Vivo

A quasispecies dynamics in vivo has been documented by following genetic variations in the course of persistent infections established with biological or molecular clones of viruses. In an early study by Narayan et al (1977) a persistent infection was established in sheep by inoculating a biological clone of the lentivirus visna virus. Antibodies were raised in the infected animals. While early antibodies neutralized the virus recovered from animals at early stages of persistence, they did not neutralize the virus recovered after one year of persistence. The result implied that mutations had occurred in visna virus that was no longer susceptible to antibody neutralization. Late antibodies, however, were able to neutralize both early and late virus. Subsequent work amply confirmed, for this and other lentiviruses, the de novo occurrence of mutations during the course of persistent infections (Clements et al, 1980; Salinovich et al, 1986; Huso and Narayan, 1990; Hammond et al, 2000).

Another early study involved the establishment of persistent FMDV infections in cattle employing biological clones of the virus (Costa Giomi et al, 1984; Gebauer et al, 1988). Genomic nucleotide sequences were determined for the virus shed by the animals for up to 539 days of persistence. Sequential genetic and antigenic changes, as well as evidence of intraisolate heterogeneity, were documented. Rates of evolution approached 10^{-1} substitutions per nucleotide position per year, an observation which influenced Temin (1989) to state that "HIV is different but not unique".

Viruses that depict the more extensive variations in vivo—such as the human hepatitis C and human (or simian) immunodeficiency viruses—generate heterogeneous collections of genomes and often rapid evolution upon inoculation of biological or molecular clones into animal hosts (Table 6.3 summarizes several experiments). In the case of hepatitis C virus, it is increasingly apparent that high mutant spectrum complexity predicts failure of response to treatment (with interferon α and the antiviral nucleoside analogue ribavirine) in infections with some viral subtypes (Pawlotsky et al, 1998). Also, the transition to chronicity as the outcome of an acute infection with hepatitis C virus has been correlated with evolution of the quasispecies replicating in the patients (Farci et al, 2000). Therefore, current evidence implicates quasispecies complexity and dynamics in viral disease progression and response to treatments.

Rapid Generation of Variant Genomes in Cell Culture

The first solid evidence for the rapid generation of viral mutants and their competitive rating was obtained with a number of viral systems in culture pioneered by the work of C. Weissmann and colleagues with bacteriophage Qβ (see also Chapters 2, 3 and 5 for a description of Qβ replicase and quantitative measurements of variant RNA replication in vitro). These early studies were instrumental in quantitating mutation rates and generation of genetic heterogeneity. That genetic heterogeneity of bacteriophage Qβ could be rather copious was suspected from the work of R. C. Valentine on the proportion of temperature sensitive mutants in phage stocks (Chapter 2). The critical development in Zürich in the early 1970s was the design of a site-directed mutagenesis procedure that took advantage of the ability of Qβ replicase to synthesize minus strand RNA in a step-wise fashion. By excluding one (or more, as needed) types of nucleoside triphosphate substrates, minus strand synthesis would proceed to a defined position of the growing RNA product where a mutagenic base analogue could then be inserted. The completed, modified minus strands served as a template for the synthesis of plus strands, some of which would include a mutation at the preselected site where the mutagenic base had been incorporated. This procedure represented the birth of reverse genetics and an early application of site-directed mutagenesis that later became the essential tool in genetics used routinely today (Weber et al, 1979). A mutant of bacteriophage Qβ with an A → G transition at position 40 from the 3' end of the genomic RNA was constructed (Domingo et al, 1976). Upon infection of *Escherichia coli*, several clones of the mutant virus reverted to the parental, wild-type sequence. In growth-competition experiments, the wild type virus displayed a clear selective advantage over the A → G mutant. By following the proportion of mutant and wild type sequences in a number of reversion and competition experiments, the rate of the specific transition G → A was estimated to be about 10^{-4} per genome doubling (Batschelet et al, 1976). This estimate of a mutation rate, together with the observation of a dynamic mutant spectrum upon passage of Qβ clones in *E. coli*, led the Zürich group to propose that: "A Qβ phage population is in a dynamic equilibrium, with viable mutants arising at a high rate on the one hand, and being strongly selected against on the other. The genome of Qβ phage cannot be described as a defined unique structure, but rather as a weighted average of a large number of different individual sequences" (Domingo et al, 1978). It is noteworthy than when individual biological clones of the variant bacteriophage Qβ were subjected to growth-competition experiments with the uncloned "wild type" population from which they were derived, they all showed a selective disadvantage (Domingo et al, 1978). The fitness of the viral quasispecies as a whole was thus higher than the fitness of most of its individual components.

Early evidence also involved detailed studies with VSV in the course of cytolytic and persistent infections in cell culture. Persistent infections of BHK-21 cells, established with a cloned stock of infectious VSV, together with defective-interfering (DI) particles, were maintained for over five years (Holland et al, 1979). Multiple mutations, far in excess of those observed for the same virus in cytolytic or acute infections in vivo, were observed. Both the standard genomes and DIs underwent continuous change during persistence, affecting several viral proteins (Rowlands et al, 1980). Rapid generation of heterogeneity and multiple types of genetic lesions were extensively characterized in cytolytic and persistent VSV populations (Spindler et al, 1982; Holland et al, 1982; O'Hara et al, 1984a,b; Steinhauer et al, 1989). Two important conclusions reached with both Qβ and VSV must be emphasized: extensive genetic heterogeneity can be attained with either no (or very limited) change in the consensus sequence of the evolving genome population (Steinhauer et al, 1989). A second observation concerns fitness of individual clones relative to the fitness of parental, average viral population. Measurement of the relative fitness of 98 individual clones, all derived from a clonal population of VSV, indicated a normal distribution of fitness values. Again, the majority of the individual clones were less fit than the original highly fit parental clonal population as a whole (Duarte et al, 1994). These observations emphasize two important conclusions pertaining to quasispecies dynamics: that genetic heterogeneity need not be accompanied by a change in the average sequence, and that there is an indeterminacy when trying to mimic the evolutionary behavior of a viral quasispecies with individual representatives of its mutant spectrum; a viral quasispecies behaves as a unit of selection (Eigen and Biebricher, 1988).

Both generation of genetic heterogeneity and a gradual change in the average nucleotide sequence were observed upon passage of biological clones of FMDV in cytolytic infections in cell culture (Sobrino et al, 1983). Independent viral lineages accumulated distinct mutations. Variation in the average sequences reflected deviation from population equilibrium, probably as a consequence of the two employed natural isolates of FMDV having a very brief history of adaptation to cell culture (Sobrino et al, 1983). Each passaged population that was tested was genetically heterogenous, with an average of two to eight mutations per infectious genome, relative to the consensus sequence of each parental FMDV population.

Generation of heterogeneous viral populations from a single infectious particle, and their evolution with serial passage, have now been documented with a large number of viruses in cell culture or in vivo, adding up to overwhelming evidence for a quasispecies dynamics in RNA replication systems (reviews in Domingo et al, 1988, 1999; Holland, 1992; Morse, 1994; Gibbs et al, 1995).

Connection between Genotype and Phenotype. Viral Quasispecies as Reservoirs of Phenotypic Variants: Episodic Selection

The relationship between genotype and phenotype (Chapter 1) has engendered continuous interest among evolutionary biologists. RNA viruses, because of their limited complexity, offer an opportunity to examine connections between genotype and phenotype, in particular to study how genetic variation relates to variation in biological traits that can be tested experimentally. This problem has been approached on theoretical grounds by mapping variations in primary sequence of simple RNA replicons with predicted secondary structures (Schuster and Stadler, 1999) (see Chapter 5). Theoretical studies have distinguished between a sequence space of neutral mutants and a sequence space of phenotypic variants (Huynen et al, 1996). It is the violation of the threshold into this second space that induces replication into error catastrophe (Chapters 4 and 8). This mapping of sequence space into a "phenotypic space" is far more complex and unpredictable when dealing with real viruses, but recent results allow some tentative proposals.

Table 6.5. Examples of viral clones isolated from mutant spectra and which show altered biological properties[a]

Antibody- or cytotoxic cell-escape mutants
Antiviral inhibitor-resistant mutants
Temperature sensitive clones
Enhanced or diminished expression (effect on translation through IRES mutations, etc.)
Altered ability to induce interferon
Increased or decreased virulence
Resistant to interfering particles
Altered cell, tissue and host tropism

[a] See text for specific references.

Two extreme scenarios could be imagined regarding the effect that mutations in viral genomes may have on recognizable biological properties of a virus. In one of the scenarios, none of the mutants of a representative mutant spectrum would show any recognizable biological alteration (in replicative properties, host range, virulence, antigenicity, etc.). In this case we should speak of a very ample space of neutral mutants among components of the quasispecies. In the second, opposite scenario, each individual variant that one could sample in a reasonable time would show some recognizable biological alteration. In this case we should speak of a very limited (or even nonexistent) space of neutral mutants among components of the quasispecies. But there is a problem: we are likely to find biological alterations that simply reflect a slightly deleterious effect of mutations in a given environment. Mutations that are merely slightly debilitating should not, in this context, be considered as phenotypically relevant variants. These types of mutants are subjected to weak negative selection (Chapter 7), and this is clearly contemplated by the neutral theory of molecular evolution (Kimura, 1983). Positive selection drives a virus to dominance because of advantageous mutations while negative selection acts to eliminate unfit viruses. These two types of selection are discussed in more detail in Chapter 7. As emphasized in a previous report (Domingo and Holland, 1994), the frequently fuzzy distinction between negative and positive selection is probably the root of the neutralist-selectionist controversy as mirrored in studies with viruses. With the exclusion of mutants that merely show minor fitness loss, which of the two scenarios, ample or limited neutrality with regard to phenotype, is closest to that which is found experimentally? Experience favors the second scenario. Table 6.5 summarizes some examples of genetic variants isolated from viral quasispecies which manifest clearly recognizable, relevant, biological alterations.

Two examples are noteworthy. One is the presence in the HIV-1 quasispecies replicating in infected individuals of mutants resistant to antiviral inhibitors, even in individuals that had not been treated with the relevant inhibitors (Nájera et al, 1995; Lech et al, 1996). The presence of inhibitor-resistant mutants is due to the high mutation frequencies within the *pol* gene of HIV-1, rendering inevitable the occasional occurrence of replacements related to drug resistance. Although the interpretation of this finding has been questioned as artefacts introduced during the RT-PCR procedures employed (Smith et al, 1997), the reality is that inhibitor-resistant HIV-1 mutants from untreated patients were not only detected by sequencing but were also isolated biologically (Nájera et al, 1995), and this has now been confirmed in several laboratories (Lech et al, 1996; Havlir et al, 1996; Quiñones-Mateu et al, 1998; review in Domingo et al, 1997). These findings are also in agreement with theoretical predictions (Coffin, 1995; Ribeiro et al, 1998). It must be stressed that it was not possible to attribute the presence of inhibitor-resistant HIV-1 mutants to transmission of resistant strains from patients subjected

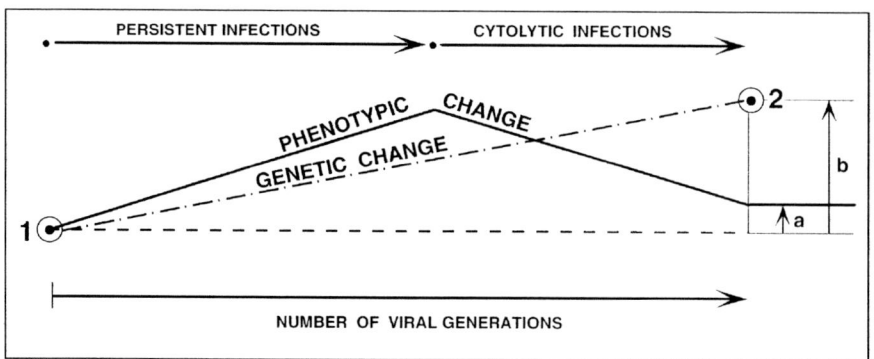

Fig. 6.5. Lack of parallelism between genetic and phenotypic change. Scheme based on the data of Sevilla and Domingo (1996) with FMDV. The transition from persistent to cytolytic infections in the same cells led to reversion of some phenotypic traits with continuing genetic diversification. At the end of this history of infections, the genetic differences between virus 2 and 1 amounted to b, while the measured phenotypic differences amounted to a (a<b).

to therapy since some of the relevant inhibitors were not in use at the time at which the HIV-1 samples were obtained (Nájera et al, 1995). The transmission of drug-resistant HIV-1 variants selected in patients subjected to antiretroviral therapy increased as a result of the widespread use of therapies with one or two inhibitors. This is referred to as a primary resistance problem, affecting a number of newly infected patients, a topic of utmost clinical importance also for multidrug-resistant bacteria and pathogenic monocellular parasites (Chapter 7).

The conclusion from a number of examples listed in Table 6.5 is that quasispecies constitute reservoirs of phenotypically relevant variants. Their adaptive value will not be manifested unless and until the replicating virus encounters the relevant selective constraint. Since selection will occur only episodically when the virus faces the appropriate environment, such events are called episodic selection, as first proposed by Dykhuizen and Hartl (1980) for bacterial polymorphic enzymes. Mutations that are just tolerated and are nearly-neutral under one set of environmental conditions may become selectively favored in another environment. The adaptive value of mutants generated during the evolution of viral quasispecies in vivo has been experimentally documented by studying phenotypes of viral clones isolated in the course of infections with human hepatitis C virus (Weiner et al, 1995; Forns et al, 1999), simian immunodeficiency virus (Kimata et al 1999; Evans et al, 1999) the arenavirus lymphocytic choriomeningitis virus (Ciurea et al, 2000), and human immunodeficiency virus (McMichael, 1998; Borrow and Shaw, 1998).

An example of partial reversion of phenotypic traits with continuous genetic diversification was described by Sevilla and Domingo (1996) with FMDV. They observed that when a virus that had persisted in cells for many cell division generations were passaged cytolytically in a same cell type, a number of phenotypic traits reverted while the virus increased its genetic distance with respect to the initial clone (Fig. 6.5). This observation illustrates that a particular phenotypic trait of a virus can be attained at different points in the genotypic sequence space. The results also established a difference between reversible and irreversible (very low probability of regaining the original trait) phenotypic modifications, suggesting that phenotypic traits may differ greatly in their robustness (insensitivity to modifications by genetic alterations) (Sevilla and Domingo, 1996).

Thresholds for Phenotypic Expression

The complexity of the mutant spectrum of a viral quasispecies has a number of implications for the expression of phenotypic traits present in viral subpopulations. The first evidence of a "buffering" capacity of a quasispecies for the dominance of a specific variant was provided by the vesicular stomatitis virus (VSV) system. A high fitness VSV clone could become dominant in an evolving VSV quasispecies only when seeded above a critical proportion in the parental VSV population (de la Torre and Holland, 1990). In a similar line of research, but using lymphocytic choriomeningitis virus (LCMV) in vivo, it was shown that the expression of a disease syndrome brought about by the virus can also depend on the proportion of pathogenic variants present in the quasispecies. Some LCMV isolates induce a growth hormone deficiency in some strains of newborn mice, while other isolates do not. When mixtures of the two types of LCMVs at different ratios were administered to mice, development of the hormone-deficiency syndrome could be prevented by an excess of the nonpathogenic variants, in spite of the pathogenic LCMV remaining at detectable levels in the infecting virus (Teng et al, 1996).

Of particular practical interest have been the studies of Chumakov and his colleagues with attenuated poliovirus vaccines. Because the transition from attenuated to virulent poliovirus is associated with one or a few mutations in the viral genome, virulent variants are expected to be present in the mutant spectrum of seed stocks and working stocks of attenuated poliovirus vaccine populations. Chumakov et al (1991) confirmed the presence of virulent variants in live poliovirus vaccines and documented that unless the virulent variants were present above a critical threshold proportion, they did not produce neurological disease in monkeys. These examples point to two facets out of a number of important biological implications of genetic microheterogeneity in viral populations. One refers to the unpredictable disease outcome when a viral infection is started by an inoculum containing mixtures of variants with different pathogenic potential. Another concerns laboratory preparations of viruses to be administered as live attenuated vaccines. Vaccinologists must be aware of the potential dangers of passaging heterogeneous viral mixtures. Although heterogeneity may be minimized by generating live virus preparations by transfection of cells with infectious RNA transcribed from a defined cDNA clone, any replicative (or transcriptional) event in the transfected cell may entail some degree of heterogeneity among progeny genomes. The presence of virulent or other types of phenotypic variants in live vaccine preparations cannot be completely ruled out with the available diagnostic procedures. Depending on the proportion and selective value of a variant type in the mutant spectrum of a quasispecies, an atypical phenotype may either be expressed or suppressed during infection.

The experimental evidence for the existence of thresholds for phenotypic expression of variant RNA viruses supports a number of theoretical arguments that imply that the fate of any specific component of the mutant spectrum of a viral quasispecies may be strongly dependent upon the total mutant spectrum which surrounds it (reviewed in Eigen and Biebricher, 1988). Again, a viral quasispecies acts as a unit of selection, and its behavior cannot be accurately predicted from the behavior of its individual components.

Host Range Mutants of RNA Viruses

Adaptability of most viruses depends on their ability to infect a range of different cell types rather than a single cell type in a highly specific fashion. Viruses can be selected either to optimize their multiplication in a certain host cell or even to change their host cell specificity. Virologists use routinely the ability of viruses to adapt to new host cells when they grow a natural isolate in embryonated chicken eggs or in established cell lines. A look at the catalogue of cell lines susceptible to different viruses reveals that many such cells are unrelated to the natural host for that virus. The human HeLa cell line can be infected by some strains of foot-and-mouth disease virus (FMDV), a virus which in nature infects mainly artiodactyls and rarely infects humans. Adaptation of viruses to unusual hosts, be they animals or cell lines, sometimes leads to variant forms of the viruses which are attenuated for their natural hosts. This is the basis of the preparation of live-attenuated vaccines, generally the most successful vaccine type for prevention of viral disease (Chapter 8). Numerous host-range mutants of both DNA and RNA viruses have been isolated (Coen and Ramig, 1996).

Representatives of the same virus group need not use the same receptor molecule to interact with the cell and initiate the infectious cycle (Evans and Almond, 1998; Schneider-Schaulier, 2000). Out of the 102 serologically distinct rhinoviruses, 90% use the so called major rhinovirus receptor or intracellular adhesion molecule-1 (ICAM-1), and 10% use the minor low density lipoprotein receptor LDLR protein, and still another rhinovirus type uses sialic acid. Picornaviruses use at least nine types of cellular receptor proteins, most of them having a cellular role related to immune responses (Evans and Almond, 1998; Rossmann et al, 2000). Interestingly, there is no obvious correlation between phylogenetic relatedness and the type of receptor used by the picornaviruses.

Changes in cell receptor or coreceptor specificity have been documented with several viruses in cell culture and in vivo. One example is provided by FMDV, an important animal pathogen with a clear potential to establish itself as a human pathogen. Natural FMDV isolates utilize as their main receptor a molecule of the ubiquitous integrin family. When FMDV is adapted to grow in cell culture, typically baby hamster kidney (BHK) cells, variant viruses able to bind heparin and to bind cells via heparan sulfate are selected (Jackson et al, 1996). This modification parallels a change in host range reflected in the acquisition of the ability to infect a new cell type, Chinese hamster ovary (CHO) cells. Mutant CHO cells defective in heparan sulfate biosynthesis are available, and they have been instrumental in showing that heparan sulfate is needed for infection by most variant FMDVs adapted to cell culture. It is not clear whether these modifications in host range are restricted to cells in culture or may also be relevant to FMDV infections in vivo. A number of alternative capsid sites can mediate heparin binding, and it is known that mutants with altered receptor specificity are present in the quasispecies of FMDV which manifest standard receptor recognition (Sa-Carvalho et al, 1997; Baranowski et al, 1998). Some FMDV variants are able to replicate efficiently in some cell lines despite not using an integrin- or heparin sulfate-dependent pathway for cell entry, suggesting that these variants may use a third mechanism for cell recognition and entry (Baranowski et al, 2000). Furthermore, mutants selected from FMDV quasispecies for their inability to bind heparin sulfate regained ability to enter cells via integrin, suggesting that this virus has the potential to shift receptor usage for entry even into the same cell type (Baranowski et al, 2000). It is not yet known what may be the significance in vivo of the capacity of FMDV to change preferential use of some receptors over others as a result of minimal modifications in this capsid.

Studies with avian retroviruses have shown that small changes in the surface glycoprotein can alter receptor usage, in response to appropriate selective pressures (Taplitz and Coffin, 1997). The human immunodeficiency viruses (HIV) probably offer the most detailed documentation of the implications of shifts in coreceptor usage in the evolution of the infection in human hosts. The number of human cells in which HIV has been found in infected individuals is impressive (Levy, 1998). It includes hematopoietic, brain, bowel and skin cells. HIV uses the cell surface antigen CD44 as its main receptor, and one of several chemokine receptors, in particular, to infect lymphocytes and macrophages, the two main target cells in the human body. In the early, asymptomatic phase of the infection, the HIV-1s that are most frequently isolated use mainly coreceptor CCR5 which mediates infection of macrophages and T cells. These early isolates do not induce syncytia (nonsyncytium-inducing, NSI phenotype) and display relatively slow rates of replication. However, HIV-1 quasispecies undergo continuous

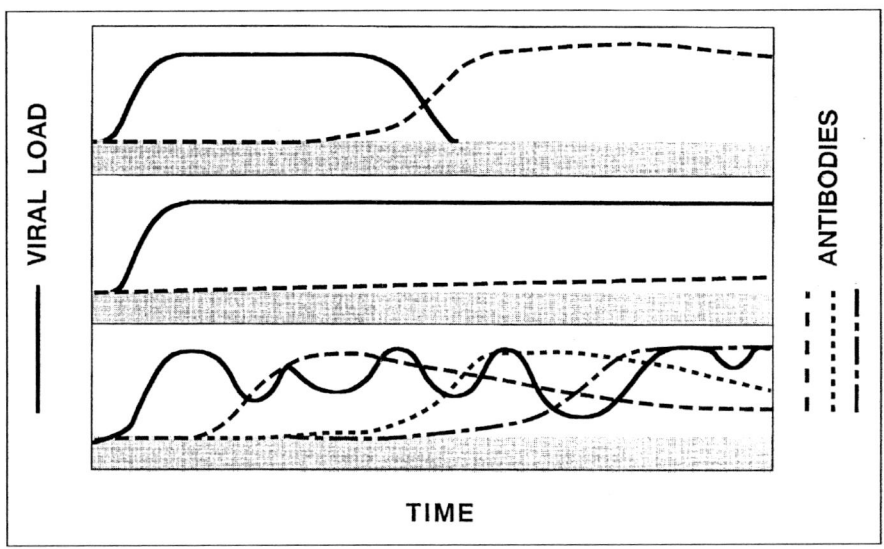

Fig. 6.6. Complexity of the effect of antibodies on an evolving viral quasispecies. Top: an appropriate antibody response (and for some virus a cytotoxic response) may lead to clearing of infection (viral load becomes undetectable). Middle: A poor antibody response may lead to viral persistence. Bottom: Fluctuations in viral load may result from antigenic variation and waves of antibody molecules evoked by viral variants (see text).

evolution in infected patients, and viruses with increasing replication rates and with a syncytium-forming (SI) phenotype in cell culture gradually become dominant. These viruses use a different type of coreceptor (CXCR4 or CXCR4 along with CCR5), and show a strong tropism for T cell lymphocytes (Berger et al, 1998; Levy, 1998). Thus, evolution in coreceptor usage is one of many important determinants of HIV-1 pathogenesis (Schneider-Schaulier, 2000).

Receptor-ligand analogues may provide a means to limit viral infection, but they may promote selection of host-range viral mutants. A modified RANTES (a natural ligand for CCR5) selected HIV-1 mutants able to use CXCR4 as a coreceptor in an in vivo mouse model (Mosier et al, 1999). The modification was facilitated by the HIV-1 isolate used, which required only one or two amino acid substitutions in its surface glycoprotein to recognise CXCR4 as a coreceptor.

Viruses may use one or several receptor molecules. They may also use a coreceptor in either an essential or dispensable manner. The same basic receptor molecule may, in association with alternative coreceptors, direct a virus to recognize one cell type or another. Viruses may also interact with cell surface molecules as landing pads prior to receptor and coreceptor usage. Receptors, coreceptors and landing pad molecules may be used efficiently or inefficiently. Low-level entry of viruses by inefficient coreceptors are increasingly documented as possible elements in disease manifestations. Shifts in recognition of high- or low-efficiency receptors, coreceptors and landing pad molecules are, at least in some cases, susceptible to modification by amino acid replacements within easy reach of the mutant spectra of viral quasispecies (Hsu et al, 1998; Baranowski et al, 1998, 2000; Hoffman et al, 1998).

Quasispecies Dynamics and Antigenic Variation

Antibodies and cytotoxic T lymphocytes (CTL) are important components of the protective responses against pathogenic agents, including viruses. As a result of antigenic variation of a pathogenic agent, the types of antibodies induced will often change. The new antibodies may no longer recognize the initial form of the pathogen, and, in turn, the variant pathogen may no longer be inactivated (neutralized) by the first set of antibodies (Fig. 6.6). Antigenic variation is critical for the survival of many RNA viruses over prolonged time periods in the field (during successive epidemic waves as for example with influenza) and also within infected individuals, in the course of a disease episode (Weiner et al, 1995; Evans et al, 1999; Ciurea et al, 2000). Perhaps the most classical example of the latter is provided by the lentiviruses equine infectious anemia virus and visna virus (reviews in Clements et al, 1988; Carpenter et al, 1990; Hammond et al, 2000). A theme of considerable debate is whether antigenic variation of viruses within infected individuals plays an important role in virus survival and disease pathogenesis or is simply a secondary phenomenon, unavoidable due to the high genetic variability of RNA viruses, but with little biological relevance. Amino acid substitutions at T cell epitopes of HIV-1 appear to have contributed to virus escape from CTLs (Koenig et al, 1995; Borrow et al, 1997; Price et al, 1997; Borrow and Shaw, 1998; McMichael, 1998; Menéndez-Arias et al, 1999). On occasions escape mutants become dominant only after many years of stability of the initial T cell epitope (Goulder et al, 1997). However, late escape of one specific epitope may be due to adverse effects on fitness of the amino acid substitution in a particular genomic sequence context, but it may still contribute to viral persistence.

The initial infection with HIV-1 results in an increase in viral load (measured as the amount of viral RNA quantitated in blood) followed within days by a decrease in viral load to a level which varies greatly from one infected individual to another. Positive correlations have been observed between CTL levels and low viral loads, a feature which is associated with the asymptomatic period of infection (Pantaleo and Fauci, 1996). However, it has been argued that the increase in CTLs coincides with the decrease in viral load but is not necessarily its direct cause (Feinberg and McLean, 1997).

Not only for HIV-1, but also for many other viruses, it has been proposed that antigenic variation may be promoted by mechanisms other than positive immune selection. According to the most generally-accepted, Darwinian model, genome variation affecting antigenic proteins occurs at random but antigenic variants do not become dominant until immune responses provide positive selection of those viruses with antigenically variant proteins. Viruses with an invariant antigenic structure would be partially or completely neutralized by the immune response (that is, their fitness relative to the antigenic variants will become very low; see Chapter 7). However, a number of observations both in vivo and in cell culture suggest that in some cases positive immune selection need not operate to raise antigenic variants to dominance (reviewed in Domingo et al, 1993). The examples affect DNA viruses as well as RNA viruses. An early study by Hampar and Keehn (1967) documented that herpes simplex virus modified its antigenic specificity in the course of a persistent infection in cell culture in the absence of antibodies. A similar observation with FMDV in persistent as well as cytolytic infections was reported independently by two groups (Bolwell et al, 1989; Díez et al, 1989). Sevilla et al, (1996) documented that two adjacent amino acid substitutions at the major antigenic site of FMDV, which abolished the reactivity of the virus with antibodies, became dominant upon serial cytolytic passage of the virus in the absence of antibodies. Interestingly, the dominance of the virus with the double amino acid substitution occurred only when large viral populations were used in the serial infections, suggesting an effect of the viral population size on the repertoire of mutants that may become dominant during quasispecies evolution. This result was particularly interesting because the two antigenically drastic replacements rose to dominance under a

Experimental Studies on Viral Quasispecies

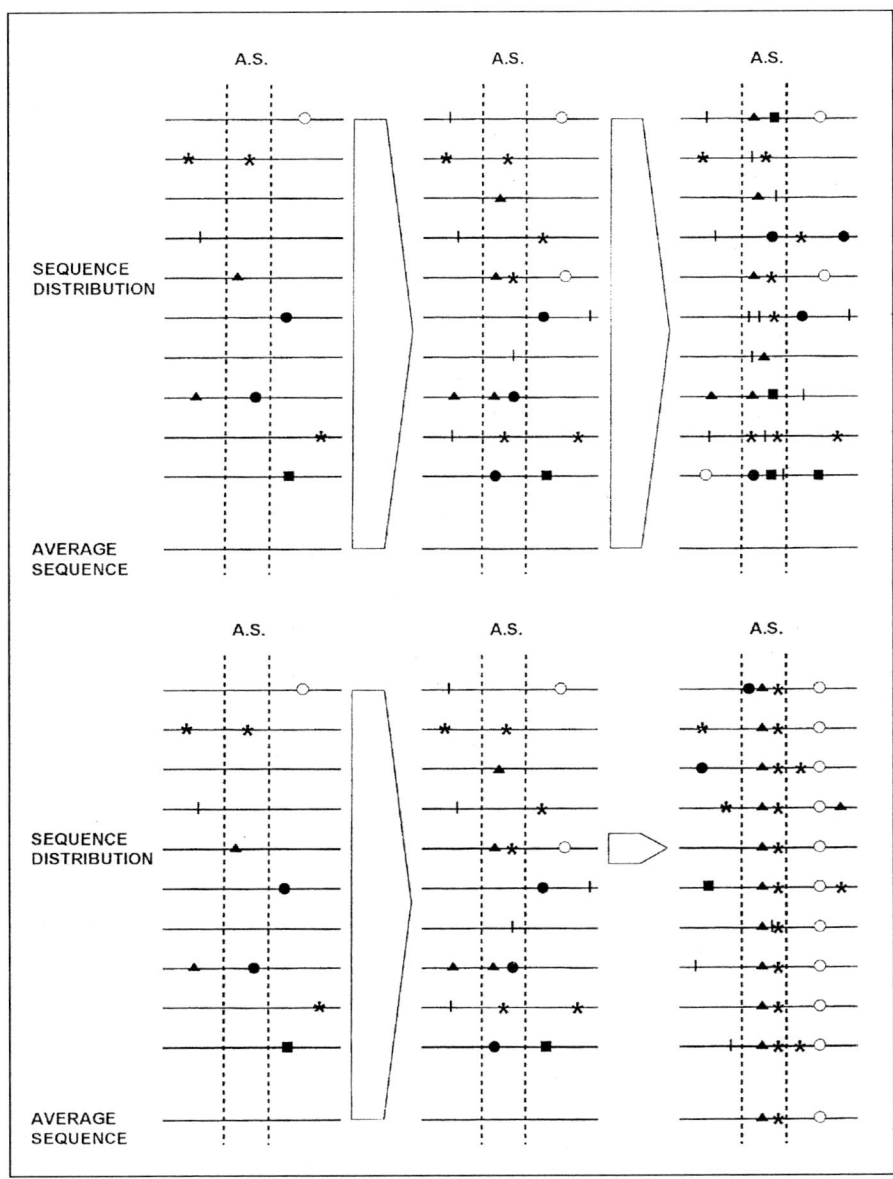

Fig. 6.7. The random change ("hitchhiking") model of antigenic variation of RNA viruses. Quasispecies are schematically depicted as in Fig. 6.4. A.S. means a region encoding an antigenic site on the viral capsid where mutations are assumed to be more tolerated than average. On top the unrestricted viral replication (big arrows) does not lead to a change in the average sequence although the mutant spectrum may show higher mutation frequencies within A.S. At the bottom, the introduction of a bottleneck (small arrow) leads to dominance of an antigenic variant reflected in the average sequence. The model need not be applied only to antigenic sites, but to any genomic region where mutations are more tolerated than average. Based in Domingo et al (1993), with permission from the Society for General Microbiology, U.K.

passage regime that minimized genetic drift (Sevilla and Domigo, 1996). It is worth noting that the repertoire of antigenic variants recorded at a defined antigenic site need not be identical in the presence and absence of antibody selection (Borrego et al, 1993).

Other viruses for which antigenic variation in the absence of immune selection has been reported include hepatitis A virus, the lentiviruses HIV-1 and equine infectious anemia virus, Newcastle disease virus, rabies virus and human and swine influenza viruses [details of the various biological contexts in which the observations were made are given in Domingo et al (1993), and references therein]. In an instructive study, Rocha et al (1991) analyzed the hemagglutinin and the nucleoprotein of sequential samples of influenza virus from a persistently infected child afflicted with a severe immunodeficiency syndrome. Antigenic variants emerged at high rates in spite of a weakened (or absent) immune response.

There are at least two main models to explain antigenic variation in the absence of an immune selection. One is that antigenic sites perform functions other than interacting with components of the immune response. For CTL responses this could be argued on the basis that T-cell epitopes are often located on conserved regions of structural or nonstructural viral proteins. In the case of B-cell epitopes, the picture has changed radically in recent years. Taking the picornaviruses as an example, it had been postulated that receptor-recognition sites were located in canyons or pits, hidden from the external surface of the particle, and protected from antibody attack. Studies with FMDV, poliovirus and rhinovirus have shown that receptor-recognition sites and antigenic sites may share one or several amino acid residues. In the case of FMDV, structural studies of complexes between peptide antigens and the Fab fragment of several antibodies have shown that amino acids which are critical for recognition of an integrin receptor are also critical for interaction with neutralizing antibodies. Antigenic sites may coincide almost exactly with receptor-recognition sites. Therefore the adaptation of a virus to a different host cell may entail an antigenic change, and antigenic variation may affect host cell recognition specificity (Verdaguer et al, 1998; Ruiz-Jarabo et al, 1999).

The second model to explain antigenic variation in the absence of immune selection is closely connected with the quasispecies structure of RNA viruses and relates to a classical concept of population biology called "hitchhiking" of mutations. The model was termed "random change model" (Domingo et al, 1993), and it is schematically depicted in Fig. 6.7. It proposes that since antigenic sites are located at the surface of viral particles they are subjected to less stringent structural constraints than internal protein domains subjected to interactions with other viral proteins or with other sites of the same protein. Obviously surface residues will never be completely free to tolerate any type of substitution (and less so when part of the antigenic domain may also be involved in receptor recognition, as in the picornaviruses discussed above). However, the model assumes that the tendency will be towards greater acceptance of substitutions at surface residues. This is supported by the location of amino acid replacements due to mutations occurring upon viral passage (in the absence of antibodies) or when comparing related isolates of the same virus. Thus, fluctuations in mutant distributions promoted by selection at unrelated loci or by founder events (as a result of genetic bottlenecks) have a higher probability of yielding dominant genomes carrying mutations at genomic loci where they are most tolerated, in particular those associated with antigenic variation (see Fig. 6.7). This mechanism necessitates the action of negative selection acting continuously on newly arising genomes. This model was originally explicitly proposed by Domingo et al (1993) and reformulated in more quantitative terms by Haydon and Woolhouse (1998). In addition to underlining the largely unpredictable nature of antigenic variation, the model also emphasizes one of the main problems encountered in studies of RNA virus evolution: the difficulties in assigning to specific mutations the phenotypic changes observed in viruses. This is because sites at which mutations are more tolerated than at average sites need not be restricted to antigenic sites, and those sites may vary by "hitchhiking" of mutations due to selection acting on unrelated loci.

References

1. Abe K, Inchauspe G, Fujisawa K. Genetic characterization and mutation rate of hepatitis C virus isolated from a patient who contracted hepatitis during an epidemic of nonA, nonB hepatitis in Japan. J Gen Virol 1992; 73:2725-29.
2. Altmuller A, Fitch WM, Scholtissek C. Biological and genetic evolution of the nucleoprotein gene of human influenza A viruses. J Gen Virol 1989; 70:2111-19.
3. Anderson JA, Bowman EH, Hu W-S. Retroviral recombination rates do not increase linearly with marker distance and are limited by the size of the recombining subpopulation. J Viol 1998; 72:1195-1202.
4. Banner LR, Lai MMC. Random nature of coronavirus RNA recombination in the absence of selection pressure. Virology 1991; 185:441-445.
5. Baranowski E, Sevilla N, Verdaguer N et al. Multiple virulence determinants of foot-and-mouth disease virus in cell culture. J Virol 1998; 72:6362-72.
6. Baranowski E, Ruiz-Jarabo CM, Sevilla N et al. Cell recognition by foot-and-mouth disease virus that lacks the RGD integrin-binding motif: Flexibility in aphthovirus receptor usage. J Virol 2000; 74:1641-47.
7. Batschelet E, Domingo E, Weissmann C. The proportion of revertant and mutant phage in a growing population, as a function of mutation and growth rate. Gene 1976; 1:27-32.
8. Benmansour A, Brahimi M, Tuffereau C et al. Rapid sequence evolution of street rabies glycoprotein is related to the highly heterogeneous nature of the viral population. Virology 1992; 187: 33-45.
9. Benmansour A, Basurco B, Monnier AF et al. Sequence variation of the glycoprotein gene identifies three distinct lineages within field isolates of viral haemorrhagic speticemia virus, a fish rhabdovirus. J Gen Virol 1997; 78:2837-46.
10. Berger EA, Doms RW, Fenys EM et al. A new classification for HIV-1. Nature 1998; 391:240.
11. Biebricher CK, Luce R. In vitro recombination and terminal elongation of RNA by $Q\beta$ replicase. EMBO J 1992; 11:5129-35.
12. Bolwell C, Brown AL, Barnett PV et al. Host cell selection of antigenic variants of foot-and-mouth disease virus. J Gen Virol 1989; 70:45-57.
13. Borrego B, Novella IS, Giralt E et al. Distinct repertoire of antigenic variants of foot-and-mouth disease virus in the presence or absence of immune selection. J Virol 1993; 67:6071-6079.
14. Borrow P, Lewicki H, Wei XP et al. Antiviral pressure exerted by HIV-1 specific cytotoxic T lymphocytes (CTLs) during primary infection demonstrated by rapid selection of CTL escape virus. Nature Med 1997; 3:205-211.
15. Borrow P, Shaw GM. Cytotoxic T-lymphoctye escape viral variants: How important are they in viral evasion of immune clearance in vivo? Immunol Rev 1998; 164:37-51.
16. Braun MJ, Clements JE, Gonda MA. The visna virus genome: evidence for a hypervariable site in the *env* gene and sequence homology among lentivirus envelope proteins. J Virol 1987; 61: 4046-54.
17. Britten RJ. Rates of DNA sequence evolution differ between taxonomic groups. Science 1986; 231:1393-98.
18. Burns DPW, Desrosiers RC. Selection of genetic variants of simian immunodeficiency virus in persistently infected rhesus monkeys. J Virol 1991; 65:1843-54.
19. Carpenter S, Evans LH, Sevoian M et al. In vivo and in vitro selection of equine infectious anemia virus variants. Applied Virology Res 1990; 2:99-115.
20. Carrillo C, Plana J, Mascarella R et al. Genetic and phenotypic variability during replication of foot-and-mouth disease virus in swine. Virology 1990; 179:890-892.
21. Cattaneo R, Billeter MA. Mutation and A/I hypermutations in measles virus persistent infections. Curr Topics in Microbiol and Immunol 1992; 176:63-74.
22. Chen Z, Li K, Rowland RR et al. Lactate dehydrogenase-elevating virus variants: cosegregation of neuropathogenicity and impaired capability for high viremic persistent infection. J Neurovirol 1998; 4:560-68.
23. Chetverin AB, Chetverina HV, Demidenko AA et al. Nonhomologous RNA recombination in a cell-free system: Evidence for a transesterification mechanism guided by secondary structure. Cell 1997; 88:503-513.

24. Chumakov KM, Powers LB, Noonan KE et al. Correlation between amount of virus with altered nucleotide sequence and the monkey test for acceptance of oral poliovaccine. Proc Natl Acad Sci USA 1991; 88:199-203.
25. Ciurea A, Klenerman P, Hunziker L et al. Viral persistence in vivo through selection of neutralizing antibody-escape variants. Proc Natl Acad Sci USA 2000; 97:2749-54.
26. Clements JE, Pederson FS, Narayan O et al. Genomic changes associated with antigenic variation of visna virus during persistent infection. Proc Natl Acad Sci USA 1980; 77:4454-58.
27. Clements JE, Gdovin SL, Montelaro RC et al. Antigenic variation in lentiviral diseases. Annu Rev Immunol 1988; 6:139-159.
28. Clewley JP, Bishop DHL, Kang CY et al. Oligonucleotide fingerprints of RNA species obtaind from rhabdoviruses belonging to the vesicular stomatitis subgroup. J Virol 1977; 23:152-166.
29. Coen DM, Ramig RF. Viral genetics. In: Fields BN, Knipe DM, Howley PM et al, ed. Fields Virology. Philadelphia: Lippincott-Raven 1996:133-151.
30. Coffin JM. HIV population dynamics in vivo: implications for genetic variation, pathogenesis and therapy. Science 1995; 267:483-489.
31. Costa Giomi MP, Bergmann IE, Scodeller EA et al. Heterogeneity of the polyribocytidylic acid tract in aphthovirus: biochemical and biological studies of viruses carrying polyribocytidylic acid tracts of different lengths. J Virol 1984; 51:799-805.
32. de la Torre JC, Holland JJ. RNA virus quasispecies can suppress vastly superior mutant progeny. J Virol 1990; 64:6278-81.
33. Díez J, Mateu MG, Domingo E. Selection of antigenic variants of foot-and-mouth disease virus in the absence of antibodies, as revealed by an in situ assay. J Gen Virol 1989; 70:3281-89.
34. Domingo E. Biological significance of viral quasispecies. Viral Hepatitis Reviews 1996; 2:247-261.
35. Domingo E. RNA virus quasispecies and their significance for viral pathogenesis. Bioforum International 1998; 2:14-16.
36. Domingo E, Flavell RA, Weissmann C. In vitro site-directed mutagenesis: Generation and properites of an infectious extracistronic mutant of bacteriophage Qβ. Gene 1976; 1:3-25.
37. Domingo E, Sabo DL, Taniguchi T et al. Nucleotide sequence heterogeneity of an RNA phage population. Cell 1978; 13:735-744.
38. Domingo E, Dávila M, Ortín J. Nucleotide sequence heterogeneity of the RNA from a natural population of foot-and-mouth disease virus. Gene 1980; 11:333-346.
39. Domingo E, Martínez-Salas E, Sobrino F et al. The quasispecies (extremely heterogeneous) nature of viral RNA genome populations: Biological relevance: A review. Gene 1985; 40:1-8.
40. Domingo E, Holland JJ. High error rates, population equilibrium, and evolution of RNA replication systems. In: Domingo E, Holland J, Ahlquist P, eds. RNA Genetics. Boca Raton, Florida: CRC Press Inc. 1988; 3:3-36.
41. Domingo E, Holland JJ, Ahlquist P, eds. RNA Genetics, vol. 1, 2, 3. Boca Raton, Florida: CRC Press Inc. 1988.
42. Domingo E, Webster RG, Holland JJ, eds. Origin and Evolution of Viruses. London: Academic Press 1999.
43. Domingo E, Mateu MG, Martínez MA et al. Genetic variability and antigenic diversity of foot-and-mouth disease virus. In: Kurstak E, Marusyk RG, Murphy SA et al, eds. Applied Virology Research, vol. II, Virus variation and epidemiology. New York: Plenum Publishing Co. 1990:233-266.
44. Domingo E, Holland JJ. Complications of RNA heterogeneity for the engineering of virus vaccines and antiviral agents. In: Setlow JK, ed. Genetic Engineering, Principles and Methods. New York: Plenum Press 1992; 14:13-32.
45. Domingo E, Díez J, Martínez MA et al. New observations on antigenic diversification of RNA viruses. Antigen variation is not dependent on immune selection. J Gen Virol 1993; 74:2039-45.
46. Domingo E, Holland JJ. Mutation rates and rapid evolution of RNA viruses. In: Morse SS, ed. Evolutionary Biology of viruses. New York: Raven Press 1994:161-184.
47. Domingo E, Holland JJ, Biebricher C et al. Quasispecies: The concept and the word. In: Gibbs A, Calisher C, García-Arenal F, eds. Molecular Basis of Virus Evolution. Cambridge University Press 1995:171-180.

48. Domingo E, Menéndez-Arias L, Quiñones-Mateu ME et al. Viral quasispecies and the problem of vaccine-escape and drug-resistant mutants. Progress in Drug Research 1997; 48:99-128.
49. Drake JW. A constant of spontaneous mutation in DNA-based microbes. Proc Natl Acad Sci USA 1991; 88:7160-7164.
50. Drake JW. Rates of spontaneous mutations among RNA viruses. Proc Natl Acad Sci USA 1993; 90:4171-4175.
51. Drake JW, Holland JJ. Mutation rates among RNA viruses. Proc Natl acad Sci USA 1999; 96:13910-13.
52. Drake JW, Charlesworth B, Charlesworth D et al. Rates of spontaneous mutation. Genetics 1998; 148:1667-86.
53. Duarte EA, Novella I, Ledesma S et al. Subclonal components of consensus fitness in an RNA virus clone. J Virol 1994; 68:4295-4301.
54. Dykhuizen D, Hartl DL. Selective neutrality of GPGD allozymes in *E. coli* and the effects of genetic background. Genetics 1980; 96:801-817.
55. Eigen M, Biebricher CK. Sequence space and quasispecies distribution. In: Domingo E, Holland JJ, Ahlquist P, eds. RNA Genetics, vol. 3. Boca Raton, Fla: CRC Press Inc. 1988; 211-245.
56. Eigen M, Nieselt-Struve K. How old is the immunodeficiency virus? AIDS 1990; 4:S85-S93.
57. Evans DJ, Almond JW. Cell receptors for picornaviruses as determinants of cell tropism and pathogenesis. Trends in Microbiol 1998; 6:198-202.
58. Evans DT, O'Connor DH, Jing P et al. Virus-specific cytotoxic T-lymphocyte responses select for amino-acid variation in simian immunodeficiency virus Env and Nef. Nature Med 1999; 5:1270-76.
59. Farci P, Schimoda A, Coiana A et al. The outcome of acute hepatitis C predicted by the evolution of the viral quasispecies. Science 2000; 288:339-344.
60. Feinberg MB, McLean AR. AIDS. Decline and fall of immune surveillance? Current Biol 1997; 7:R136-R140.
61. Fitch WM, Bush RM, Bender CA et al. Long term trends in the evolution of H(3) HAI human influenza type A. Proc Natl Acad Sci USA 1997; 94:7712-18.
62. Forms X, Purcell RH, Bukh J. Quasispecies in viral persistence and pathogenesis of hepatitis C virus. Trends Microbiol 1999; 7:402-710.
63. Gebauer F, de la Torre JC, Gomes I et al. Rapid selection of genetic and antigenic variants of foot-and-mouth disease virus during persistence in cattle. J Virol 1988; 62:2041-49.
64. Gessain A, Gallo RC, Franchini G. Low degree of human T-cell leukemia/lymphoma virus type I genetic drift in vivo as a means of monitoring viral transmission and movement of ancient human populations. J Virol 1992; 66:2288-95.
65. Gibbs A, Calisher CH, García-Arenal, F, eds. Molecular Basis of Virus Evolution. Cambridge: Cambridge University Press, 1995.
66. Giménez-Barcons M, Ibañez A, Tajahuerce A et al. Genetic evolution of hepatitis G virus in chronically infected individual patients. J Gen Virol 1998; 79:2623-29.
67. Gojobori T, Yokoyama S. Rates of evolution of the retroviral oncogene of Maloney murine sarcoma virus and of its cellular homologues. Proc Natl Acad Sci USA 1985; 82:4198-4201.
68. Gorman OT, Bean WJ, Kawaoka Y et al. Evolution of the nucleoprotein gene of influenza A virus. J Virol 1990; 64:1487-97.
69. Goulder PJR, Phillips RE, Colbert RA et al. Late escape from an immunodominant cytotoxic T-lymphocyte response associated with progression to AIDS. Nature Med 1997; 3:212-217.
70. Hahn BH, Shaw GM, Taylor ME et al. Genetic variation in HTLV-III/LA over time in patients with AIDS or at risk for AIDS. Science 1986; 232:1548-53.
71. Hammond SA, Li F, McKeon BM et al. Immune responses and viral replication in long-term inapparent carrier ponies inoculated with equine infectious anemia virus. J Virol 2000; 74:5968-81.
72. Hampar B, Keehn MA. Cumulative changes in the antigenic properties of herpes simplex virus from persistently infected cell cultures. J Immunol 1967; 99:554-7.
73. Havlir DV, Eastman S, Gamst A et al. Nevirapine-resistant human immunodeficiency virus: kinetics of replication and estimated prevalence in untreated patients. J Virol 1996; 70:7894-99.
74. Haydon DT, Woolhouse ME. Immune avoidance strategies in RNA viruses: Fitness continuums arising from trade-offs between immunogenicity and antigenic variability. J Theor Biol 1998; 193:601-612.

75. Hoffman TL, Stephens EB, Narayan O et al. HIV type 1 envelope determinants for use of the CCR2b, CCR3, STRL33, and APJ coreceptors. Proc Natl Acad Sci USA 1998; 95:11360-365.
76. Holland JJ ed. Genetic Diversity of RNA viruses. Curr Top Microb Immunol, vol. 176. Berlin: Springer-Verlag, 1992.
77. Holland JJ, Grabau EA, Jones CL et al. Evolution of multiple genome mutations during long-term persistent infection of vesicular stomatitis virus. Cell 1979; 16:495-504.
78. Holland JJ, Spindler K, Horodyski F et al. Rapid evolution of RNA genomes. Science 1982; 215:1577-85.
79. Huelsenbeck JP, Rannala B. Phylogenetic methods come of age: Testing hypothesis in an evolutionary context. Science 1997; 276:227-232.
80. Hsu EC, Sarangi F, Iorio C et al. A single amino acid change in the hemagglutinin protein of measles virus determines its ability to bind CD46 and reveals another receptor on marmoset B cells. J Virol 1998; 72:2905-16.
81. Huso DL, Narayan O. Escape of lentiviruses from immune surveillance. Applied Virology Res 1990; 2:61-73.
82. Huynen MA, Stadler PF, Fontana W. Smoothness within ruggedness: the role of neutrality in adaptation. Proc Natl Acad Sci USA 1996; 93:397-401.
83. Imazeki F, Omata M, Ohto M. Heterogeneity and evolution rates of delta virus RNA sequences. J Virol 1990; 64:5594-95.
84. Ito T, Gorman OT, Kawaoka Y et al. Evolutionary analysis of the influenza A virus M gene with comparison of the M1 and M2 proteins. J Virol 1991; 65:5491-98.
85. Jackson T, Ellard FM, Abu Ghazaleh R et al. Efficient infection of cells in culture by type O foot-and-mouth disease virus requires binding to cell surface heparan sulfate. J Virol 1996; 70: 5282-87.
86. Kew OM, Nottay BK, RicoHesse R et al. Molecular epidemiology of wild poliovirus transmission. Applied Virol Res 1990; 2:100-221.
87. Kew OM, Mulders MN, Lipskaya GY et al. Molecular epidemiology of polioviruses. Seminars in Virol 1995; 6:401-414.
88. Kew OM, Sutter RW, Nottay BK et al. Prolonged replication of a type 1 vaccine-derived poliovirus in an immunodeficient patient. J Clin Microbiol 1998; 36:2893-99.
89. Khatchikian D, Orlich M, Rott R. Increased viral pathogenicity after insertion of a 28S ribosomal RNA sequence into the hemagglutinin gene of an influenza virus. Nature 1989; 340:156-157.
90. Kimata JT, Kuller L, Anderson DB et al. Emerging cytopathic and antigenic simian immunodeficiency virus variants influence AIDS progression. Nature Med 1999; 5:535-541.
91. Kimura M. The neutral theory of molecular evolution. Cambridge: Cambridge University Press; 1983.
92. Kirkegaard K, Baltimore D. The mechanism of RNA recombination in poliovirus. Cell 1986; 47: 433-443.
93. Koenig S, Conley AJ, Brewah YA et al. Transfer of HIV-1-specific cytotoxic T lymphocytes to an AIDS patient leads to selection of mutant HIV variants and subsequent disease progression. Nat Med 1995; 1:330-336.
94. Kuge S, Kawamura N, Nomoto A. Strong inclination toward transition mutation in nucleotide substitutions by poliovirus replicase. J Mol Biol 1989; 207:175-182.
95. Kümmerer BM, Stoll D, Meyers G. Bovine viral diarrhea virus strain Oregon: a novel mechanism for processing of NS2-3 based on point mutations. J Virol 1998; 72:4127-38.
96. Kurath G, Rey MEC, Dodds JA. Analysis of genetic heterogeneity within the type strain of satellite tobacco mosaic virus reveals several variants and a strong bias for G to A substitution mutation. Virology 1992; 189:233-244.
97. Lai MMC. Recombination and its evolutionary effect on viruses with RNA genomes. In: Gibbs A, Calisher CH, García-Arenal, eds. Molecular basis of virus evolution. Cambridge: Cambridge University Press; 1995:119-132.
98. Lech WJ, Wang G, Yang YL et al. In vivo sequence diversity of the protease of human immunodeficiency virus type 1: Presence of protease inhibitor-resistant variants in untreated subjects. J Virol 1996; 70:2038-43.

99. Lee C-M, Bih F-Y, Chao Y-C et al. Evolution of hepatitis delta virus RNA during chronic infection. Viology 1992; 188:265-273.
100. Leung DW, Chen E, Goeddel DV. A method for random mutagenesis of a defined DNA segment using a modified polymerase chain reaction. Technique 1989; 1:11-15.
101. Levy JA. HIV and the pathogenesis of AIDS. Washington DC: ASM Press, 1998.
102. Luytjes W, Bredenbeek PJ, Noten AF et al. Sequence of mouse hepatitis virus A59 mRNA2: Indications for RNA recombination between coronaviruses and influenza C virus. Virology 1988; 166: 415-422.
103. Mansky LM. The mutation rate of human immunodeficiency virus type 1 is influenced by the *vpr* gene. Virology 1996; 222:391-400.
104. Martínez MA, Dopazo J, Hernández J et al. Evolution of the capsid protein genes of foot-and-mouth disease virus: Antigenic variation without accumulation of amino acid subtitutions over six decades. J Virol 1992; 66:3557-65.
105. Martínez MA, Vartanian J-P, Wain-Hobson S. Exploring the functional robustness of an enzyme by in vitro evolution. EMBO J 1996; 15:1203-10.
106. McMichael A. T cell responses and viral escape. Cell 1998; 93:673-676.
107. Menéndez-Arias L, Mas A, Domingo E. Cytotoxic T lymphocyte response to HIV-1 reverse transcripase. Viral Immunol 1999; 11:167-181.
108. Meyerhans A, Vartanian J-P. The fidelity of cellular and viral polymerases and its manipulation for hypermutagenesis. In: Domingo E, Webster RG, Holland JJ, eds. Origin and Evolution of Viruses. San Diego: Academic Press 1999:87-114.
109. Meyers G, Rumenapf, Thiel HJ. Ubiquitin in a togavirus. Nature 1989; 341:491.
110. Minor PD. The molecular biology of poliovaccines. J Gen Virol 1992; 73:3065-77.
111. Monroe SS, Schlesinger S. RNAs from two independently isolated defective interfering particles of Sindbis virus contain cellular tRNA sequences at their 5' ends. Proc Natl Acad Sci USA 1983; 80: 3279-83.
112. Morimoto K, Hooper DC, Cargaugh H et al. Rabies virus quasispecies: Implications for pathogenesis. Proc Natl Acad Sci USA 1998; 95:3152-56.
113. Morse SS, ed. The Evolutionary Biology of Viruses. New York: Raven Press, 1994.
114. Mosier DE, Picchio GR, Gulizia RJ et al. Highly potent RANTES analogues either prevent CCR5-using human immunodeficiency virus type 1 infection in vivo or rapidly select for CXCR4-using variants. J Virol 1999; 73:3544-50.
115. Nagy PD, Simon AE. New insights into the mechanisms of RNA recombination. Virology 1997; 235:1-9.
116. Nájera I, Holguín A, Quiñones-Mateu ME. The *pol* gene quasispecies of human immunodeficiency virus. Mutations associated with drug resistance in virus from patients undergoing no drug therapy. J Virol 1995; 69:23-31.
117. NaKao H, Okamoto H, Fukuda M et al. Mutation rate of GB virus C/hepatitis G virus over the entire genome and in subgenomic regions. Virology 1997; 233:43-50.
118. Narayan O, Griffin DE, Chase J. Antigenic drift of visna virus in persistently infected sheep. Science 1977; 197:376-378.
119. Ogata N, Alter HJ, Miller RH et al. Nucleotide sequence and mutation rate of the H strain of hepatitis C virus. Proc Natl Acad Sci USA 1991; 88:3392-96.
120. O'Hara PJ, Horodyski FM, Nichol ST et al. Vesicular stomatitis virus mutants resistant to defective-interfering particles accumulate stable 5'-terminal and fewer 3'-terminal mutations in a stepwise manner. J Virol 1984; 49:793-798.
121. O'Hara PJ, Nichol ST, Horodyski FM et al. Vesicular stomatitis virus defective interfering particles can contain extensive genome sequence rearrangements and base substitutions. Cell 1984b; 36: 915-924.
122. Okamoto H, Kojima M, Okada S-I et al. Genetic drift of hepatitis C virus during a 8.2-year infection in a chimpanzee: variability and stability. Virology 1992; 190:894-899.
123. Ortín J, Nájera R, López C et al. Genetic variability of Hong Kong (H3N2) influenza viruses: spontaneous mutations and their location in the viral genome. Gene 1980; 11:319-331.
124. Page RDM, Holmes EC. Molecular Evolution. A phylogenetic approach. Oxford: Blackwell Science Ltd. 1998.

125. Pantaleo G, Fauci AS. Immunopathogenesis of HIV infection. Annu Rev Immunol 1996; 50: 825-854.
126. Parrish CR, Aquadro CF, Strassheim ML et al. Rapid antigenic-type replacement and DNA sequence evolution of canine parvovirus. J Virol 1991; 65:6544-52.
127. Pawlotsky J-M, Germanidis G, Neumann AU et al. Interferon resistance of hepatitis C virus genotype 1b: Relationship to nonstructural 5A gene quasispecies mutations. J Virol 1998; 72:2795-2805.
128. Pickering JM, Thomas HC, Karayannis P. Genetic diversity between hepatitis G virus isolates: Analysis of nucleotide variation in the NS-3 and putative "core" peptide genes. J Gen Virol 1997; 78:53-60.
129. Price DA, Goulder PJR, Klenerman P et al. Positive selection of HIV-1 cytotoxic T lymphocyte escape variants during primary infection. Proc Natl Acad Sci USA 1997; 94:1890-95.
130. Querat G, Audoly G, Sonigo P et al. Nucleotide sequence analysis of SA-OMVV, a Visna-related ovine lentivirus: phylogenetic history of lentiviruses. Virology 1990; 175:434-447.
131. Quiñones-Mateu ME, Albright JL, Mas A et al. Analysis of *pol* gene heterogeneity, viral quasispecies, and drug resistance in individuals infected with group O strains of human immunodeficiency virus type 1. J Virol 1998; 72:9002-15.
132. Ribeiro RM, Bonhoeffer S, Nowak MA. The frequency of resistant mutant virus before antiviral therapy. AIDS 1998; 12:461-465.
133. Rocha E, Cox NJ, Black RA et al. Antigenic and genetic variation in influenza A (H1N1) virus isolates recovered from a persistently infected immunodeficient child. J Virol 1991; 65:2340-50.
134. Rossman MG, Bell J, Kolatkar PR et al. Cell recognition and entry by rhino- and enteroviruses. Virology 2000; 269:239-247.
135. Rota PA, Wallis TR, Harmon MW et al. Cocirculation of two distinct evolutionary lineages of influenza type B virus since 1983. Virology 1990; 175:59-68.
136. Rowe CL, Fleming JO, Nathan MJ et al. Generation of coronavirus spike deletion variants by high-frequency recombination at regions of predicted RNA secondary structure. J Virol 1997; 71: 6183-90.
137. Rowlands D, Grabau EA, Spindler K et al. Virus protein changes and RNA termini alterations evolving during persistent infection. Cell 1980; 19:871-880.
138. Rueda P, García Barreno B, Melero JA. Loss of conserved cysteine residues in the attachment glycoprotein of two human respiratory syncytial virus escape mutants that contain multiple A-G substitutions (hypermutation). Virology 1994; 198:653-662.
139. Ruiz-Jarabo C, Sevilla, N, Dávila M et al. Antigenic properties and population stability of a foot-and-mouth disease virus with an altered Arg-Gly-Asp receptor-recognition motif. J Gen Virol 1999; 80:1899-1909.
140. Rupprecht CE, Smith JS, Fekadu M et al. The ascension of wild type rabies: a cause for public health concern or intervention? Emerg Infect Dis 1995; 1:107-114.
141. Sa-Carvalho D, Rieder E, Baxt B et al. Tissue culture adaptation of foot-and-mouth disease virus selects viruses that bind to heparin and are attenuated in cattle. J Virol 1997; 71:5115-23.
142. Salemi M, Vandamme A-M, Gradozzi C et al. Evolutionary rate and genetic heterogeneity of human T-cell lymphotropic virus type II (HTLV-II) using isolates from European injecting drug users. J Mol Evol 1998; 46:602-611.
143. Salinovich O, Paune SL, Montelaro RC et al. Rapid emergence of novel antigenic and genetic variants of equine infectious anemia virus during persistent infection. J Virol 1986; 57:71-80.
144. Schneider-Schaulier J. Cellular receptors for viruses: Links to tropism and pathogenesis. J Gen Virol 2000; 81:1413-29.
145. Schrag SJ, Rota PA, Bellini WJ. Spontaneous mutation rate of measles virus: Direct estimation based on mutations conferring monoclonal antibody resistance. J Virol 1999; 73:51-54.
146. Schuster P, Stadler PF. Nature and evolution of early replicons. In: Domingo E, Webster RG, Holland JJ, eds. Origin and Evolution of Viruses. London: Academic Press, 1999.
147. Sedivy JM, Capone JP, Raj Bhandary UL et al. An inducible mammalian amber suppressor: propagation of a poliovirus mutant. Cell 1987; 50:379-389.
148. Sevilla N, Domingo E. Evolution of a persistent aphthovirus in cytolytic infections: partial reversion of phenotypic traits accompanied by genetic diversification. J Virol 1996; 70:6617-6624.

149. Sevilla N, Verdaguer N, Domingo E. Antigenically profound amino acid substitutions occur during large population passages of foot-and-mouth disease virus. Virology 1996; 225:400-405.
150. Sharp PM, Robertson DL, Gao F, Hahn BH. Origins and diversity of human immunodeficiency viruses. AIDS 1994; 8:S27-S42.
151. Smith DB, Pathirana S, Davidson F, Lawlor E, Power J, Yap PL, Simmonds P. The origin of hepatitis C virus genotypes. J Gen Virol 1997; 78:321-328.
152. Smith DB, McAllister J, Casino C et al. Virus "quasispecies": making a mountain out of a molehill? J Gen Virol 1997; 78:1511-19.
153. Smith FI, Palese P. Influenza viruses: high rate of mutation and evolution. In: Domingo E, Holland JJ, Ahlquist P, eds. RNA Genetics, vol. 3. Boca Raton: CRC Press 1988; 123-135.
154. Smith TF, Srinivasan A, Schochetman G et al. The phylogenetic history of immunodeficiency viruses. Nature 1988; 333:573-575.
155. Sobrino F, Dávila M, Ortín J et al. Multiple genetic variants arise in the course of replication of foot-and-mouth disease virus in cell culture. Virology 1983; 128:310-318.
156. Sobrino F, Palma EL, Beck E et al. Fixation of mutations in the viral genome during an outbreak of foot-and-mouth disease: heterogeneity and rate variations. Gene 1986; 50:149-159.
157. Spindler KR, Horodyski FM, Holland JJ. High multiplicities of infection favor rapid and random evolution of vesicular stomatitis virus. Virology 1982; 119:96-108.
158. Steinhauer DA, de la Torre JC, Meier E et al. Extreme heterogeneity in populations of vesicular stomatitis virus. J Virol 1989; 63:2072-80.
159. Stech J, Xiong X, Scholtissek C et al. Independence of evolutionary and mutational rates after transmission of avian influenza viruses to swine. J Virol 1999; 73:1878-84.
160. Strauss JH. Recombination in the evolution of RNA viruses. In: Morse SS, ed. Emerging viruses. New York: Oxford University Press 1993:241-251.
161. Sugimoto C, Kitamura T, Guo J et al. Typing of urinary JC virus DNA offers a novel means of tracing human migrations. Proc Natl Acad Sci USA 1997; 94:9191-96.
162. Suzuki Y, Yamaguchi-Kabata Y, Gojobori T. Nucleotide substitution rates of HIV-1. AIDS Rev 2000; 2:39-47.
163. Takeda N, Tanimura M, Miyamura K. Molecular evolution of the major capsid protein VP1 of enterovirus 70. J Virol 1994; 68:854-862.
164. Taplitz RA, Coffin JM. Selection of an avian retrovirus mutant with extended receptor usage. J Virol 1997; 71:7814-19.
165. Temin H. Is HIV unique or merely different? J AIDS 1989; 2:1-9.
166. Teng MN, Oldstone MBA, de la Torre JC. Suppression of lymphocytic choriomeningitis virus-induced growth hormone deficiency syndrome by disease-negative virus variants. Virology 1996; 223:111-119.
167. Toja M, Escarmís C, Domingo E. Genomic nucleotide sequence of a foot-and-mouth disease virus clone and its persistent derivatives. Implications for the evolution of viral quasispecies during a persistent infection. Virus Res 1999; in press.
168. Van Ranst M, Kaplan JB, Sundbery JP et al. Molecular evolution of papillomaviruses. In: Gibbs A, Calisher CH, Garcia-Arenal F, eds. Molecular Basis of Virus Evolution. Cambridge: Cambridge University Press 1995; 455-476.
169. Vartanian J-P, Henry M, Wain-Hobson S. Hypermutagenic PCR involving all four transitions and a sizeable proportion of transversions. Nucleic Acids Res 1996; 24:2627-31.
170. Vartanian J-P, Meyerhans A, Asjo B et al. Selection, recombination and G × A hypermutation of human immunodeficiency virus type 1 genomes. J Virol 1991; 65:1779-88.
171. Verdaguer N, Sevilla N, Valero ML et al. A similar pattern of interaction for different antibodies with a major antigenic site of foot-and-mouth disease virus: Implications for intratypic antigenic variation. J Virol 1998; 72:739-748.
172. Viazov S, Riffelmann M, Khoudyakov Y et al. Genetic heterogeneity of hepatitis G virus isolates from different parts of the world. J Gen Virol 1997; 78:577-581.
173. Villaverde A, Martínez MA, Sobrino F et al. Fixation of mutations at the VP1 gene of foot-and-mouth disease virus. Can quasispecies define a transient molecular clock? Gene 1991; 103:147-153.
174. Wain-Hobson S. Running the gamut of retroviral variation. Trends in Microbiol 1996; 4:135-141.

175. Wattel E, Vartanian J-P, Pannetier C, Wain-Hobson S. Clonal expansion of human T-cell leukemia virus type I-infected cells in asymptomatic and symptomatic carriers without malignancy. J Virol 1995; 69:2863-68.
176. Weaver S, Scot TW, RicoHesse R. Molecular evolution of eastern equine encephalomyelitis virus in North America. Virology 1991; 182:774-784.
177. Weber H, Taniguchi T, Müller W et al. Application of site-directed mutagenesis to RNA and DNA genomes. Cold Spring Harbor Symp Quant Biol 1979; 43:669-677.
178. Webster RG, Bean WJ, Gorman OT et al. Evolution and ecology of influenza A viruses. Microbiol Rev 1995; 56:152-179.
179. Weiner A, Erickson AL, Kansopon J et al. Persistent hepatitis C virus infection in a chimpanzee is associated with emergence of a cytotoxic T lymphocyte escape variant. Proc Natl Acad Sci USA 1955; 92:2755-59.
180. Weissmann C, Billeter MA, Goodman HM et al. Structure and function of phage RNA. Annu Rev Biochem 1973; 42:303-328.
181. Yamashita M, Krystal M, Fitch WM et al. Influenza B virus evolution: cocirculating lineages and comparison of evolutionary pattern with those of influenza A and C viruses. Virology 1988; 163:112-22.
182. Zaccolo M, Gherardi E. The effect of high-frequency random mutagenesis on in vitro protein evolution: A study of TEM-1 β-lactamase. J Mol Biol 1999; 285:775-83.
183. Zuckerkandl E, Pauling L. Evolutionary divergence and convergence in proteins. In: Bryson V, Vogel HJ, eds. Evolving Genes and Proteins. New York: Academic Press 1965; 97-166.

CHAPTER 7

Population Dynamics and Virus Adaptability
Viral Fitness

The meaning of fitness of living organisms has evolved since the time of Darwin (Reznick and Travis, 1996). The concept has broadened from referring to the survival of an individual to meaning its lifetime reproductive success. The latter, in turn, encompasses a number of elements such as growth rate, age of maturity, frequency of reproduction, number of offspring, success of offspring and others. Maximal reproductive success cannot be measured simply by fecundity but rather by the resources that may be available to each of the offspring to ensure their own reproductive success (Williams, 1992; Villarreal, 1999).

Fitness values are relative and transient. Relative because they change with environmental parameters including the presence and activities of other organisms. Transient because the physical and biological surroundings are generally dynamic, not static. Fitness has a heritable component built into the genetic makeup of the organism, and a nonheritable component which is a composite of environmental parameters.

Equally complex is the concept of fitness as applied to viruses (Domingo and Holland, 1997). It has been very useful to adapt an operational definition of viral fitness that renders the concept amenable to relatively simple experimental designs. Viral fitness has been defined as the relative ability of a virus population to produce infectious progeny under a set of defined environmental conditions. Values are necessarily relative, and they are often determined in growth-competition experiments involving the virus to be tested and a reference "wild type" virus which is arbitrarily given a fitness value of 1 (Holland et al, 1991). The experiment consists of serial infections of host cells (or animals) with the two viruses mixed in a known proportion. The progeny of the first round of infection are used to carry out a second round of infection of the host cells (or animal) under the same experimental conditions. This process of competition passages is repeated a number of times, generally not exceeding five. The relative proportion of the two viruses is determined for the initial mixture and for the progeny of the successive passages. The proportion of the two competing viruses at each passage, relative to that in the initial mixture, is used to derive a fitness vector the slope of which is taken as the relative fitness value (Holland et al, 1991; Duarte et al, 1992). Unless the long-term evolution of the two competing viruses needs to be tested, for fitness determinations it is advisable to restrict the number of competition passages. This will minimize effects of additional mutations (unavoidable during RNA genome replication) and of variations of multiplicity of infection (the number of infectious virus per cell) on the relative growth properties of the two competing viruses (Sevilla et al, 1998).

Quasispecies and RNA Virus Evolution: Principles and Consequences, edited by Esteban Domingo, Christof K. Biebricher, Manfred Eigen and John J. Holland. ©2001 Eurekah.com.

The virus to be tested and the reference virus must be distinguished either phenotypically or genetically in order to determine accurately their relative amounts. A phenotypic trait that has been widely used is resistance to a monoclonal antibody (MAb). Depending on the nature of the antigenic site for which a neutralizing MAb is available, it may be possible to derive a neutral viral mutant that cannot be neutralized by the MAb. Then in competitions involving a MAb-sensitive and a MAb-resistant virus, the proportion of the two variants can be determined by plating samples of virus from each competition passage in the presence and in the absence of MAb. Since complete resistance to a MAb is unlikely, it is important to adjust the MAb concentration to allow >90% survival of the resistant mutant and minimal (<10%) survival of the sensitive virus. The survival frequencies will depend on the relative ability of the MAb to neutralize the two viruses, and they will determine the limits of detection of the two viruses in the competition mixtures and, therefore, the accuracy of the fitness vector. As a general rule, the MAb should be added to the agar overlay after attachment of the virus mixture, once virus penetration into the cells has taken place (Chapter 2). Incubations of the mixed viral populations with the antibody prior to plating should be avoided since mixed infections at high multiplicities of infection may produce phenotypic masking of MAb-resistant genomes which involves incorporation of MAb-sensitive proteins in their particles (Holland et al, 1989; Valcarcel and Ortín, 1989). Other phenotypic traits, such as different sensitivity to an antiviral inhibitor, plaque size or morphology may also be used for fitness determinations.

An alternative procedure increasingly used to quantitate relative fitness involves nucleotide sequence determinations at sites where the two competing viruses differ in sequence (Escarmís et al, 1999; Yuste et al, 1999). Their relative amounts can be quantitated through densitometry or image analysis techniques. Instead of direct sequencing, sampling methods of genome variation (gel mobility assays, restriction fragment length variations) have been also adapted for fitness determinations of viruses. These procedures, carried out both in cell culture and in vivo, are increasingly documenting the great adaptability of RNA viruses as reflected by adjustment of fitness values in response to environment changes.

Repeated Population Bottlenecks Lead to Fitness Losses: Muller's Ratchet

Studies with a number of different RNA viruses have documented a remarkable influence of the population size of replicating virus in fitness variation. The results constitute a clear demonstration of the quasispecies dynamics undergone by RNA viruses during their multiplication and its profound influence in virus adaptability. The first documentation that a drastic reduction in population size resulted in a fitness decrease was reported by Chao (1990). He subjected the tripartite (genome consisting of three double-stranded RNA molecules) *Pseudomonas phaseolicola* phage ϕ6 to repeated plaque-to-plaque transfers, the most severe form of bottleneck event that a virus may undergo since its effective population size is periodically reduced to one (Chapter 2). The presence of a mutant spectrum in RNA virus populations predicts that each plaque transfer has a good probability of isolating a virus deviating in one or several mutations from the average. A tendency to incorporate deleterious mutations was first proposed by Muller (1964) to operate in small populations of asexual organisms when no compensatory mechanisms such as sex or recombination intervene. Such accumulation of deleterious mutations which result in average fitness losses is known as Muller's ratchet. Its operation has been documented with RNA viruses of animals and with protozoa (Bell, 1988), the fish *Poeciliopsis monacha* (Leslie and Vrijenhoek, 1980), and bacteria (Andersson and Hughes, 1996).

The debilitating effects of plaque-to-plaque transfers of RNA viruses may be viewed as an accentuation of the ratchet effect predicted by Muller. Stochastic fitness decreases, some very intense, as a result of subjecting virus to repeated bottleneck events have been observed with the animal viruses vesicular stomatitis virus (VSV) (Duarte et al, 1992, 1993), foot-and-mouth disease virus (FMDV) (Escarmís et al, 1996) and human immunodeficiency virus type 1 (HIV-1) (Yuste et al, 1999). Yuste et al (1999) emphasized that the extent of fitness loss differed greatly among the three animal viruses subjected to a similar experimental design of serial plaque transfers. A number of HIV-1 clones were so debilitated that the virus was extinguished (drastic loss of plating efficiency) before the intended number of plaque transfers could be completed. The number of extinctions decreased for the other viruses in the order HIV-1 > FMDV > VSV (or φ6). The reasons for these differences are not well understood, but they may relate to the different numbers of replication rounds occurring for each of the viruses in the course of plaque formation. Indeed plaque growth is the only step in the process of plaque transfer that permits some competitive selection among viral genomes. The number of infectious particles found in a plaque of VSV (10^6-10^8 plaque-forming units or p.f.u.; see Fig. 2.2 in Chapter 2) is higher than found for FMDV (10^5 to 10^6 p.f.u.) and much higher than found in a plaque of HIV-1 (10^3 to 10^4 p.f.u., in the assay employed by Yuste et al, 1999). It is likely that the limitation of viral population size in the successive plaques facilitates the accumulation of deleterious mutations. The stronger accentuation of Muller's ratchet effect for HIV-1 could also be influenced by differences in mutation and recombination rates among the three viruses. However, as summarized in Chapter 6, it is unlikely that differences in mutation rates between retroviruses and riboviruses (Drake, 1993) could account for marked differences observed in fitness decrease. Also, recombination would be expected to compensate for fitness losses, and no evidence of recombination (except for formation of defective interfering particles, see Chapter 6) has been obtained for VSV or other mononegavirales.

Molecular Basis of Fitness Decrease

The molecular events underlying fitness loss as a result of plaque-to-plaque transfers were analyzed in the case of FMDV (Escarmís et al, 1996). A summary of the types of genetic lesions that accumulated in FMDV clones as a result of bottleneck events is summarized in Table 7.1. The most remarkable lesion was an internal polyadenylate of up to 35 residues that represented an extension of four adenosines that precede the second functional AUG used for initiation of protein synthesis. The genomic RNA of FMDV has two functional AUG initiation codons that direct the synthesis of two forms of a leader (L) protease that differ only at their amino-terminal region. It is likely that the internal polyadenylate was tolerated because the two forms of L are functionally redundant and the internal poly A affected the synthesis of only one form of L (Escarmís et al, 1996). Furthermore, this debilitating lesion could be isolated and sequenced because the virus harboring it did not need to compete with more fit viruses after cloning.

Half of the amino acid substitutions at the structural proteins of the debilitated FMDV clones occupied internal positions in the capsid. In contrast, when mapping capsid substitutions in a variety of FMDV lineages that had been subjected to large population passages, only 4% of the substitutions were found at internal locations in the capsid; the majority (96%) affected residues accessible to a probe of a size similar to a water molecule (Escarmís et al, 1996). Thus, there were important qualitative changes in the types of mutations found in FMDV clones subjected to repeated bottlenecks, and related FMDV populations derived from large population passages (Table 7.1). (Our view of FMDV variation had been strongly influenced by the routine passages used to produce high titer stocks!). How can such a difference in the types of mutations be explained? The interpretation is straightforward considering the presence of a mutant spectrum in the replicating population of genomes, and the events are depicted in Figure 7.1. Assume that the initial quasispecies distribution corresponds to a virus which is well adapted to the environment. The random sampling of one genome from the distribution (arrow in Fig. 7.1, that represents growth from a single infectious genome during plaque develop-

Table 7.1. Genetic lesions acompanying FMDV passages[a]

TYPE OF MUTATION	FOUND IN VIRUS SUBJECTED TO:	
	PLAQUE-TO-PLAQUE TRANSFERS	LARGE POPULATION PASSAGES
Transitions	77%	69%
Transversions	23%	31%
Nonsynonymous	49%	76%
Synonymous	51%	24%
Internal polyadenylate extension[b]	In 50% of clones	Not found[b]
Amino acid replacements internal in the capsid[c]	50%	4%

[a] Based on results with FMDV clone C-S8c1 reported by Escarmis et al (1996).
[b] The internal polyadenylate represents an extension of five adenosine residues found at positions 1119 to 1123 of the FMDV C-S8c1 genome. No such internal polyadenylate extension has been detected in about one hundred FMDV populations (biological clones or natural isolates) subjected to large population passages.
[c] Internal means not accessible to a probe of a size similar to a water molecule.

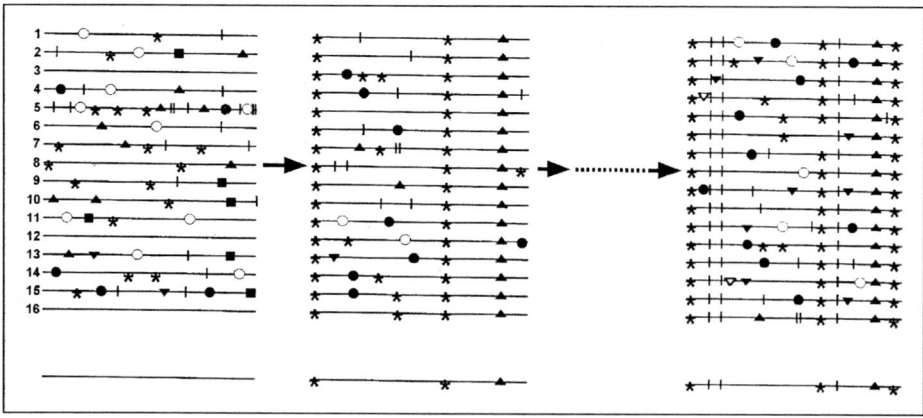

Fig. 7.1. Schematic representation of the effect of serial bottleneck events on a viral quasispecies. Genomes are indicated by horizontal lines, and symbols on the genomes represent mutations. Most genomes from the initial quasispecies (left) deviate from the master, represented here by the zero mutation class of genomes (molecules 3, 12 and 16 from the sequence distribution). Bottleneck events (arrows) lead to amplification of one genome from the distribution, resulting in gradual accumulation of mutations, with a gradual modification of the average sequence. When the initial quasispecies is well adapted to the environment, repeated bottleneck events are generally expected to result in average fitness decreases, since genomic distributions gradually deviate from the well adapted one. This prediction from the quasispecies structure of RNA viruses has been amply confirmed experimentally (see text). (from Domingo et al, 1996, with permission from FASEB J.).

ment) will tend to divert the newly generated quasispecies from its previous optimum. Successive random sampling (bottleneck) events will gradually increase the mutational load resulting in fitness decrease. The parallel accumulation of debilitating mutations and a decrease in fitness were clearly documented in the analyses with FMDV (Escarmís et al, 1996). Very few mutations scattered throughout the genome were sufficient to take the virus to a fitness minimum. An unfit FMDV clone found to be near extinction (due to Muller's ratchet) included, in addition to the internal polyadenylate discussed in preceding paragraphs, a total of seven point mutations (four synonymous or affecting noncoding regions and three affecting amino acids of nonstructural proteins). Independent clones subjected to plaque transfers in parallel lineages accumulated unique sets of mutations but in similar proportions (Escarmís et al, 1996). This suggests that under the experimental conditions employed in these experiments, the number of mutations needed for FMDV to move in the fitness landscape (Wright, 1931, 1982) is commensurate with the number present in the mutant spectra of FMDV quasispecies with a not too-distant clonal origin [7 mutations in a 8500-residue genome represents a mutation frequency of 8×10^{-4} substitutions per nucleotide, a value which characterizes the mutant spectrum complexity of many FMDV populations (reviewed in Domingo et al, 1990, 1992)].

Fitness decreases are often observed in biological clones of viruses which have been isolated after surviving a strong selective pressure directed to the bulk of the population. Examples are polyclonal antibody-resistant clones of FMDV (Borrego et al, 1993) or inhibitor-resistant HIV-1 variants (Goudsmit et al, 1996; Harrigan et al, 1998; Nijhuis et al, 1999; Zennou et al, 1998). Individual clones harboring the mutations needed to respond to the selective constraint are a minority in the quasispecies, reflecting a selective disadvantage relative to the quasispecies average (Domingo et al, 1978). These minority clones often show higher fitness than the average population when fitness is measured in the presence of the selective agent, as expected. In most cases, extensive multiplication of the virus, either in the absence or the presence of the selective agent, leads to fitness recovery.

Fitness Increase of RNA Viruses: Quasispecies Optimization

Large population passages of RNA viruses generally result in fitness gains when fitness is measured in the same physical and biological environment in which the passages were carried out (Martínez et al, 1991; Clarke et al, 1993; Novella et al, 1995a; Escarmís et al, 1999). In an extensive study with several VSV clones, Novella et al (1995a) found that fitness increases followed exponential kinetics. When the initial clone had a very low fitness, a two-phase kinetics was observed, with a rapid, initial fitness increase until neutrality was attained. Novella et al (1995a) pointed out that the spectacular fitness gain from a single viral particle denotes an adaptive evolutionary capacity that "overwhelms speculation". For one of the VSV clones analyzed, fitness increase reached 5000% after 50 passages!. As a comparison with a DNA-based organism, the fitness gain during adaptation of the bacterium *Escherichia coli* to new culture conditions was around 8% in 400 generations and 37% in 2000 generations (Bennet et al, 1990; Lenski and Travisano, 1994).

Given that large population passages ensure rapid fitness gains while serial bottleneck events lead to stochastic fitness losses, a very relevant question in the population dynamics of RNA viruses is the population size needed to maintain a constant fitness value in a given environment. Again, studies with VSV clones indicated that such critical population size depends on the initial fitness of the viral clone. The higher the initial fitness value, the less severe must be the bottleneck to maintain fitness (Novella et al, 1995b). Debilitated viral clones often gain fitness in spite of being subjected to rather severe bottlenecks (Novella et al, 1995b; Elena et al, 1998), a treatment that would have led to stochastic fitness losses had the initial fitness been high. Coherently with these results, Novella et al (1999b) have recently found that as the fitness of an evolving VSV population increases, even large population passages cannot ensure a continuing fitness gain. At this critical point of the fitness-population landscape, fitness variations become erratic, probably manifesting the effects of stochastic mutations that drive the virus to neighboring points of the fitness landscape, without a clear direction or trend. These observations pose interesting questions regarding the possible effects for virus adaptation of temporary residence of virus population in such an ambiguous position in an adaptive landscape.

Fitness variations independent of externaly applied selective pressures have been detected among FMDV clones isolated from infected swine at different times after infection (Carrillo et al, 1998). By appropriate competition experiments in vivo it was shown that a clone isolated during the early viremic state of swine maintained a replicative advantage in swine, in competition with other clones.

Simian immunodeficiency virus replicating in macaques undergo sequential genetic and antigenic variations that render the virus increasingly fit for replication and to cause disease in this primate host (Kimata et al, 1999). (Other examples of adaptive evolution of viral quasispecies are discussed in Chapter 6).

Multiple Molecular Pathways for Fitness Increase: A Wrightian View of RNA Virus Evolution

A number of altered, low fitness clones of RNA viruses have been analyzed with regard to modifications undergone by genomic RNA as replication competence is restored. An FMDV clone that had been extremely debilitated by operation of Muller's ratchet was subjected to large population passages to determine the genetic modifications associated with fitness gain (Escarmís et al, 1999). Entire genomic sequences were determined as fitness increased. By passage 100, the relative fitness of the population had increased 30-fold relative to the initial, debilitated clone. The two consensus nucleotide sequences differed at 15 sites. Only two of the 15 changes represented true reversions to the sequence of the parental clone. The remaining genetic modifications were additional mutations, some of them probably pseudo-reversions affecting coding and noncoding regions, including ten amino acid replacements at structural and nonstructural proteins. Comparison of sequential genomic sequences revealed that loss of the internal polyadenylate extension was the first genetic alteration associated with fitness gain. Interestingly, the study of fitness recovery underwent by four subclones of the same debilitated clone revealed three distinct molecular pathways for the loss of the internal polyadenylate (Escarmís et al, 1999).

Results with a number of viral systems point to similar conclusions. A broad spectrum of revertants, pseudorevertants and quasi-infectious viruses (those that generate infectious progeny different from the infecting genome) are usually isolated following transfections with molecular clones (or transcripts) of a number of viruses that had been altered by site-directed mutagenesis [RNA bacteriophages (Licis et al, 1998), picornaviruses (Wimmer et al, 1993; Agol, 1997; Gromeier et al, 1999) or HIV-1 (Berkhout et al, 1997) provide some examples]. Although in most cases relative fitness values were not determined, noninfectious or poorly infectious clones gave rise to replication-competent, modified progeny. The view emerging from these experimental results is, again, that RNA viruses are capable of generating altered genomic RNA with high frequency. These altered RNAs offer multiple choices for fitness gain via a number of alternative molecular pathways.

These observations with RNA viruses have a bearing on alternative views defended by population geneticists who constructed the so called "modern synthesis" of Mendelism and Darwinism. Fisher (1930) was a strong advocate of adaptations being largely the result of small evolutionary steps. Wright (1931, 1982) accepted a larger possible contribution of chance in evolution and the possibility of attaining alternative fitness peaks in a complex fitness landscape with many mountains, valleys, canyons and a network of optional routes to go uphill or downhill in the fitness landscape. This Wrightian view of virus adaptation has been also assumed for viruses that must alternate between hosts in their life cycles (Scott et al, 1994). These arguments, however, should not be taken as any type of provocative confrontation between Fisher and Wright (they had enough in their life times!). Indeed one of the main arguments of Fisher (1930)—that large genetic variations are more likely to be deleterious than beneficial— is widely accepted for genomes which are well adapted to their environments. In the examples of fitness gain of viruses described above, the starting genomes had been severely debilitated either by a history of bottleneck events or by site-directed mutagenesis of critical genomic regions. The picture emerging from studies with viruses might be different if the question addressed were the types of molecular events underlying further fitness gains of an already quite fit genome.

Current Views on Fitness Evolution of RNA Viruses

In spite of the fitness of RNA viruses currently being under intense investigation, and conclusions still being tentative, a few general trends can be suggested from observations summarized in preceding paragraphs.
- Fitness variations are rapid. A single passage of a virus in cell culture or in vivo may produce a major fitness change, reflecting a modification of the quasispecies distribution.
- As a result of the isolation of an individual clone from a viral quasispecies, either through random sampling or through selection, fitness of the individual clone tends to be lower than the fitness of the parental quasispecies –especially for high fitness starting populations. Passage of the clonal population in the same environment will often result in fitness increase.
- Fitness of clones isolated following application of an antiviral selective agent (antibodies, inhibitors, etc.) is often lower than fitness of the parental quasispecies, provided fitness is measured in the absence of the selective agent. When fitness is measured in the presence of the selective agent, fitness of the selected clone tends to be higher than fitness of the parental quasispecies.
- Population size modulates fitness variations. In general, infections in which the virus population size is greatly reduced will often lead to fitness losses. In contrast, infections involving transfer of large virus populations will often lead to fitness gain. However, the effect of population size is strongly dependent on the initial fitness of the population. The higher the initial fitness, the higher is the population size needed to ensure that the infection will not lead to fitness loss. Stochastic effects are much more pronounced for infections by very small virus populations.

Certainly, as more research on viral fitness is completed, more accurate and additional generalizations will become possible and desirable, since they may help both in the understanding of quasispecies dynamics and in the design of preventive and therapeutic regimes against RNA viral disease.

Complexity of Fitness Landscapes

Wright depicted adaptive landscapes in the form of mountains of high fitness and plains and valleys of low fitness. This is a convenient way to depict fitness variations of viruses since genomic changes, in combination with environmental modifications, drive viruses uphill or downhill in the fitness landscape. Fitness landscapes for viruses are complex, dynamic and highly unpredictable. In a two-dimensional version of a fitness landscape, a virus may lie on a mountain ridge along which the virus can easily move. Its chances of survival are high (Fig. 7.2).

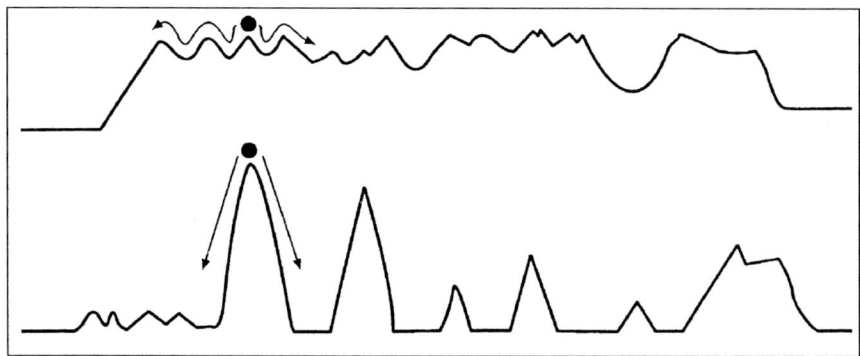

Fig. 7.2. Schematic two-dimensional version of a fitness landscape. Top: A virus population (black dot) may be located on a mountain ridge, therefore facilitating adaptation (movements). Bottom: A virus may be at the top of an isolated fitness peak, with a slim probability of reaching other fitness peaks (adaptation).

Readily observed genetic and fitness modifications among antibody-resistant and inhibitor-resistant mutants belong to this category. In another type of fitness landscape, a virus may be located at the top of an isolated fitness peak (Fig. 7.2). Perhaps such specialized high fitness might at times correspond to unique evolutionary events such as the one that led to an influenza RNA-ribosomal RNA nonhomologous recombinant found in a virulent influenza virus (see Chapter 6). No evidence of evolutionary continuity of such an influenza recombinant has been obtained, suggesting an extinction-prone genetic constellation. A virus on an isolated peak has a low chance of survival when selective conditions change. In a multidimensional fitness landscape the probability of finding ridges raises considerably.

For viruses that use alternative hosts, such as the arboviruses which alternate between mammalian and insect hosts, it is important to be able to shift hosts without detrimental fitness losses (Weaver et al, 1992; Weaver, 1998). Host alternation and virus persistence in the insect vector have been regarded as selective constraints that may promote stasis of average sequences in arboviral quasispecies. In a recent study, Weaver and colleagues have documented that eastern equine encephalitis virus (EEEV) fixed more mutations upon adaptation to either insect or mammalian cells than upon alternated passages in the two cell lines. However, passages in one cell type alone resulted in fitness gains that were not larger than those attained after alternating cell passages (Weaver et al, 1999). In this particular cell culture model system alternating host infections did appear to constrain the rates of viral evolution but not the ability of the virus to gain replicative fitness in each of the two host cell types. Using another arbovirus, Novella et al (1999) did not observe a difference in the accumulation of mutations in the VSV genome subjected to passage in either mammalian cells or insect cells alone, or in alternating passages. It is likely that a number of factors, in addition to the need to replicate in different cell types, contribute to the slow evolution often seen for arboviruses.

Evidence of Positive and Negative Selection Acting Continuously in Viral Populations

In preceding sections of this Chapter we have described passage regimes leading to fitness gains or losses, as well as some of the underlying molecular events in the viral genomes. Virus adaptation can also be explained in a slightly different way, emphasizing the difference between positive or Darwinian selection and negative or purifying selection (Kimura, 1983; Williams,

1992; Page and Holmes, 1998). Positive or Darwinian selection is the force that drives an individual or group of individuals (organisms, cells or viruses) to dominance because of advantageous mutations (or gene constellations) in a given environment. Negative or purifying selection is the force that reduces the frequency of an individual or groups of individuals because of mutations which are deleterious in a given environment. The recent literature on RNA virus evolution includes multitudes of examples of viral genomes subjected to positive or negative selection, and many are discussed in other Chapters of this book. Antibody- or cytotoxic T lymphocyte (CTL)-escape mutants, documented with several viral systems (Borrow and Shaw, 1998) may participate in viral persistence and constitute examples of positive selection acting in vivo. Similarly, inhibitor-resistant mutants of RNA viruses may be driven to dominance in an infected individual treated with the inhibitor by positive selection if no perturbing factors intervene. Consequently, inhibitors or antibodies/CTL exert negative selection upon the "wild type" sensitive viruses.

If negative selection eliminates unfit mutants, and as a consequence, unfit mutants cannot be detected in viral quasispecies, how does one know that they occurred in the first place? Several lines of experimental evidence support the occurrence of such mutants (Domingo, 1998). (i) When a strong selective agent is applied to a replicating viral quasispecies, unusual mutants, better fit to replicate in the presence of the selective agent, are frequently isolated. An example was provided by FMDV mutants that escaped neutralization by polyclonal antibodies targeted to one specific antigenic site of the viral capsid. These mutants had multiple amino acid replacements at and near the relevant antigenic site, and displayed very low replicative fitness when measured in the absence of antibodies (Borrego et al, 1993). Obviously, the selected mutants had to be generated in the course of FMDV replication since all progeny FMDVs were derived from a single genome in an initial cloning event. Their fate was to be eliminated or maintained at low frequencies by negative selection unless antibodies that imparted a selective advantage to them were present.

A different example which underlines the same concept is provided by the FMDV genomes harboring an internal polyadenylate extension (described in previous sections). This particular site on the FMDV genome (four adenylate residues preceding the second functional AUG initiation codon) would qualify as a mutational "hot spot", according to the exceedingly high frequency of clones with an increased number of adenylate residues found among clones subjected to repeated bottlenecks (Escarmís et al, 1996). Yet the absence of any observed modification of this locus in hundreds of other FMDV populations can best be explained by negative selection in spite of a tendency of the replicase to "stutter" at homopolymeric tracts (Kunkel, 1990; Bebenek et al, 1993; Escarmís et al, 1996; Meyerhans and Vartanian, 1999).

The repertoire of RNA virus variants that we see in populations passaged under standard conditions in cell culture, or in animals, as well as the genomic sequences that enter data banks and phylogenetic analyses, have all passed the filters imposed by negative selection. Hence, we see an infinitely small proportion of the myriads of variations being continuously tested in each cell infected by any RNA virus. Both negative and positive selection keep acting at the different stages of viral spread: from one cell compartment to another, from cell-to-cell, tissue-to-tissue, organ-to-organ, organism-to-organism (transmission) and also during persistence at the level of the cell, the individual or in a population of hosts (Domingo et al, 1998). Suggestions that most genetic change in RNA viruses are neutral (with no biological relevance) cannot be reconciled with the observations, now established both in vitro and in vivo, that RNA virus fitness is continuously changing—often drastically. Were most mutations neutral, a fitness change would make head-lines in highly reputed scientific journals. It is now clear that we must keep viruses in the deep-freezer to ensure invariant fitness....

Competition Between Neutral Variants: the Competitive Exclusion Principle and the Red Queen Hypothesis. Reproducible Nonlinear Population Dynamics

The analysis of the outcome of competitions between VSV clones of similar fitness has suggested the operation of a number of important principles of modern population biology during RNA virus evolution (Clarke et al, 1994). Prolonged competition between two VSV clones of nearly neutral fitness resulted in the coexistence of the two populations for many generations (passages) until one of them suddenly displaced the other. Eventually stochastic changes occurred that perturbed the initial equilibrium between the two competing populations (Clarke et al, 1994). This observation agrees with the competitive exclusion principle of population genetics which states that when species compete and share the same biological niche, one will always eventually dominate (or exclude) the other (Gause, 1971). RNA viruses may be more prone than complex organisms to sudden exclusions because of their extremely high error rates during RNA genome replication (Chapter 6). Unless the competing clones both have a high initial fitness, any advantageous mutation in one of the quasispecies may eventually arise to drive it to dominance. Cases of prolonged coexistence of competing, high fitness FMDV subpopulations have been recently described (Sevilla et al, 1998; Ruiz-Jarabo et al, 1999).

In the initially-equal competition series with VSV clones (Clarke et al, 1994), both the winner and the loser clones gained fitness in the course of passaging. Therefore, the expected fitness increase mediated by large population passages (see previous section) occurred independently of the fate (towards dominance or extinction) of the competing populations. This generalized fitness improvement whereby the relative fitness of the two competitors remains quite close, at least for a time, was formulated by Van Valen (1973) and it is known as the Red Queen hypothesis. The colorful and literary designation alludes to the words of the Red Queen in Lewis Carroll's novel *Through the Looking Glass*: "It takes all the running you can do to keep in the same place". In any competition, "no species can ever win and new adversaries grinningly replace the losers" (van Valen, 1973).

The displacement of one population by another need not always be a stochastic process subjected to an indeterminate outcome. When many replicas of two closely related, competing VSV quasispecies were followed in a constant cell culture environment, a highly predictable, nonlinear behavior of the two competing populations was found (Quer et al, 1996). Here, predictable behavior means that critical points at which competitions deviated from equilibrium (one quasispecies became dominant over the other) were reached after nearly constant periods of time (generations). These results provide evidence of some deterministic behavior, as the one attributed to quasispecies in the theoretical development of the concept (Chapter 4). Evidence of a deterministic behavior requires statistically significant amounts of data, which is difficult to realize experimentally. These observations with VSV clones are reminiscent of some predictions with nonlinear dynamic systems investigated in physics. The critical points observed in the competition among VSV clones bear resemblances to glitches (sudden noise introductions) which perturb deterministic trajectories in chaotic systems (Grebogi et al, 1990; compare with Quer et al, 1996). The nature of such glitches in viral competitions may actually be a class of perturbing mutations, but evidence for this possibility has not yet been provided. A similar mechanism may explain reproducible fluctuations among different FMDV serotypes recorded in tissue culture (Woodbury et al, 1995). A deterministic behavior of RNA viruses during natural infections of differentiated organisms is unlikely ever to be demonstrated. Individual hosts, by virtue of genetic polymorphisms, will rarely present identical environments to the virus (Hill, 1998). Differences may affect receptor molecules, subsets of partially permissive cells, immune and inflammatory responses, metabolic pathways, etc. These environmental parameters may modulate the relative abundance of quasispecies swarms, rendering unique the evolution of an RNA virus in each infected individual (Holland et al, 1992; see also Chapter 6).

Genetic Heterogeneity and Variability of DNA Viruses

The possibility that those DNA viruses that do not use RNA copies of their genomes as replicative intermediates may behave as quasispecies has been considered (Smith and Inglis, 1987; Domingo and Holland, 1994; Holland and Domingo, 1998). The limited evidence now available suggests that a distinction should be made between large, complex DNA viruses and smaller DNA genome (RNA-like) viruses. Small DNA viral genomes are often single-stranded which has a strong influence on the copying fidelity. DNA viruses are likely to differ in their tolerance to genetic variation, depending on their genetic complexity (Chapter 4). The tailed DNA bacteriophages have genome sizes that range from 20,000 to 752,000 base pairs (bp)! The parvoviruses of mammals and birds have a single stranded DNA genome of about 5000 residues, while the double stranded DNA genome of poxviruses, iridoviruses and herpesviruses can reach up to 370,000 bp. In contrast to RNA viruses, the range of genome sizes for DNA viruses spans two orders of magnitude.

In molecular epidemiological studies, genetic diversity of DNA viruses has been amply documented, even for the most complex ones. For example, restriction fragment length polymorphisms have been reported in comparing independent isolates of herpes simplex virus (HSV), and also among clones from a single virus source (Lonsdale et al, 1980; Umene et al, 1984; Sakaoka et al, 1995; Yamaguchi et al, 1998). Particularly variable appears to be the copy number of tandem-reiterated sequences found in several regions of HSV genomes. Variations at these genomic loci extend even to plaque-purified clones derived from a single isolate. The molecular basis for these variations is not well understood but it may involve polymerase slippage in copying repeated sequences (Umene et al, 1984; Yamaguchi et al, 1998). A remarkable diversity has been discovered in the gene encoding an early glycoprotein of the human herpesvirus 8 (HHV8), a virus which is associated with Kaposi's sarcoma (Zong et al, 1999). Sequence analyses of 60 different samples from four continents revealed amino acid replacements at 62% of the amino acids of this glycoprotein. The different sequences defined four subtypes and many clades. When multiple samples of the same patient were compared, the sequences were always identical. The frequency of drug-resistant mutants in clinical isolates and laboratory preparations of HSV ranged from 10^{-2}-10^{-5} (Sarisky et al, 2000).

Genomic variations in baculoviruses have been associated with recombination events affecting the length of repeat elements (García-Maruniak et al, 1996). Baculovirus heterogeneity has been revealed either by biological cloning in cell culture or by limited-dilution infection in vivo (Muñoz et al, 1999). Thus a number of complex DNA viruses often generate microheterogeneities by modifying the number of repeats in reiterated sequences.

In an attempt to quantitate point mutation frequencies, Yañez et al (1991) carried out repetitive nucleotide sequencing of a dispensable genomic segment of a clonal population of African swine fever virus. Their result established a maximum mutation frequency of 5.5×10^{-5} substitutions per nucleotide for this complex virus. A revealing result comparing variation of herpes simplex virus (HSV) in the field, with variation of human immunodeficiency virus type 1 (HIV-1), was provided by the studies of Rojas et al (1993, 1994). Comparing isolates from the same geographical area at comparable time intervals, they derived a star phylogeny (genetic lineages radiated from a single source) for both viruses. Yet the average branch length was 100-times shorter for HSV than for HIV-1! Herpes viruses have coevolved with their hosts for prolonged time periods to maintain replication functions (McGeoch and Davison, 1999) and it would be surprising if they were capable of episodes of very rapid evolution affecting most of their genes, given the profound roots of these complex viruses in the evolutionary history of their hosts (Drake, 1993; Holland and Domingo, 1998).

An intimate coevolution with their hosts is also suggested by the number of cellular homologs present in large DNA viral genomes. These viral-coded counterparts of cellular proteins manifest activity in counteracting various host defense mechanisms (Murphy, 1993; Alcamí et al, 1998; Smith et al, 1998). Some are suppressors of surface molecules such as MHC class I and class II, others are homologues of the extracellular binding domains of cytokine receptors, etc. Although host-interfering proteins are also encoded by several RNA viruses, it is among the complex DNA viruses where they appear to be widely exploited for adaptation. The counterpart of the escape-strategy of RNA virus appears to be a capture-modulation strategy in DNA viruses. These different solutions to survival probably reflect both the long-term coevolution of DNA viruses with their hosts, and their limited genetic flexibility to use high mutation rates as an effective means to overcome host immune responses and other selective constraints.

The polyomavirus SV40 has been known for a long time to exhibit considerable genetic variation, including genetic exchange with the host cells (Winocour et al, 1980). SV40 is suspected to be a human pathogen since it has been detected in some human tumors. Variants of SV40 arose de novo in simian immunodeficiency virus (SIV)-infected, immunocompromised rhesus monkeys afflicted with SV40 brain disease (Lednicky et al, 1998). In spite of great potential for variation of SV40 within individual hosts, other papovaviruses evolve comparatively slowly as they spread in the human population. Papillomaviruses isolated 30 years apart in different continents may show 99.9% DNA sequence identity, with rates of evolution as low as 3×10^{-8} substitutions per nucleotide per year (Van Ranst et al, 1995), a rate 10^5-fold slower than for typical RNA viruses (Chapter 6). In spite of this overall slow evolution, phylogenetic trees relating papillomaviruses are extensively ramified (Chan et al, 1992), and in the case of the *epidermodysplasia verruciformis* (EV) group, there is increased genetic variation to the point that mixtures of EV variants are frequently found in EV-associated lesions (Van Ranst et al, 1995). The estimated evolutionary rate of polyomavirus JC (JCV) in the human population is $1\text{-}3 \times 10^{-7}$ substitutions per nucleotide per year (Sugimoto et al, 1997). A number of JCV subtypes were distinguished and they were often found within defined geographic boundaries. Pockets with virus subtypes that differed from the prevalent subtypes in a given area were found; some of them could be explained by human migrations, since persistent, asymptomatic JCV constitutes a relatively stable genetic marker for human populations (Sugimoto et al, 1997).

Single stranded DNA viruses such as the human and animal parvoviruses may evolve at rates that approach the values for some RNA viruses, and mutation-mediated shifts in host species have been documented (Parrish and Truyen, 1999). The DNA viruses for which the most clear evidence of quasispecies dynamics has been obtained are the plant geminiviruses (Isnard et al, 1998). Geminiviruses have been grouped into species, genera and strains. Different strains show 90% to 100% nucleotide sequence identity, and clones from the same strain often show some genetic differences, as expected from a viral quasispecies-like dynamics. There is also ample evidence that interspecies recombination plays a role in geminivirus diversification, and some rapid expansions of geminiviruses towards new geographical areas appear to be associated with recombination events.

Studies with most DNA viruses have not been sufficiently thorough to allow firm conclusions regarding their involvement in a quasispecies dynamics, at least not as extensively as it has been documented for RNA viruses. However, it is clear that any simple replicon (be it an RNA or a DNA, a virus, a plasmid, a viroid or a satellite) will display a quasispecies behavior whenever copying accuracy is limited and the number of replication rounds as a function of time is large, an expected consequence of the general principles of Darwinian evolution. It may be tentatively assumed that complex DNA viruses will not generally exploit an average high mutability throughout their entire genome for adaptation due to the lethal character of such indiscriminate high mutation rates. However, loci-specific variations, such as those occurring at tandem repeats, or at specific open-reading frames or at defined recombination hot spots may occur and be exploited for adaptation. Genetic instability of triplet repeats in the human chromosome are associated with a number of genetic diseases (Wells et al, 1998). Even cells with the unquestionable high complexity in the genetic information they carry, may show features of quasispecies dynamics, since no sharp, clear discontinuities separate biological entities with regard to adaptive strategies.

Extensions to Nonviral Systems: Mutation, Competition and Selection in Cell Populations

Some types of cells, and some genetic elements within cells, share the short-term variation, competition and selection typical of RNA viruses. A number of retroid agents intimately entrenched in the life of the cell employ an error-prone reverse transcription step in their replication (McClure, 1999). A mutation rate of 2×10^{-5} substitutions per nucleotide copied has been determined for the *Sacharomyces* retrotransposon Ty1 (Gabriel et al, 1996). Retroid agents are likely to have contributed to genetic variation in cells and to their continuing adaptation. Up to 20% of the genome of a typical mammalian cell may be composed of retroelements (Baltimore, 1985; McClure, 1999). It has been suggested that episodes of general hypermutability in cellular organisms may have contributed to intervals of accelerated evolution during certain time periods (Erwin and Valentine, 1984; Agur and Kerszberg, 1987). Frequent genetic change, and selective episodes following genetic change, is a mechanism currently operating in differentiated organisms, notably in the generation of immunoglobulin diversity and in the expansion of tumor cells in metastasis.

In the course of an immune response the antibodies produced by B lymphocytes show a gradual increase in affinity for the antigen that triggered the response, in a process known as affinity maturation. This is a complex series of events, including somatic hypermutation processes of immunoglobulin genes (Berek and Milstein, 1988). Mutations are preferentially found in rearranged V-D-J sequences of immunoglobulin genes, with an increasing number of mutations as the immune response progresses. B cells expressing immunoglobulins with higher affinity for antigen are selected and expanded. Mutations occur at random (as in the mutant repertoires of RNA viruses), and only those mutations which are capable of triggering a selection event will be biologically exploited. The process of somatic hypermutation affects small segments of only one specific DNA strand and mutation rates attain values in the range of 10^{-3} substitutions per nucleotide and cell generation (Kallberg et al, 1996; Storb, 1996). Not only point mutations but also insertions, deletions, other recombination events and incorporation of nontemplate-directed nucleotides (termed "N" nucleotides) by terminal transferase have been observed. The selective alteration of only one of the DNA strands may be an evolutionary adaptation to spare unnecessary genetic damage (or biological commitment) to all progeny B cells. Hypermutations occur at germinal centers where activated B cells accumulate. Hypermutagenesis is a transient event that ceases as soon as the B cell matures. Mutations accumulate in the complementarity-determining regions (CDRs), the protein loops which interact with the epitope that triggered synthesis of the relevant immunoglobulin lineage. The molecular basis of such directed hypermutational events are poorly understood but suppression of repair-correction mechanisms or error-prone repair processes may be involved. Low fidelity DNA polymerases have been described (Matsuda et al, 2000, and references therein).

An increased mutational input, no matter how controlled and localized may it be, is not without dangers for cell survival and general stability of the intricate processes involved in an immune response. It has been estimated that up to 50% of the mutations employed for the generation of immunoglobulin diversity may lead to nonfunctional antibodies (due to a variety of defects in the molecules). The cell may have evolved molecular devices to suppress the effects of potentially lethal mutations. The "N" nucleotides (nontemplate-directed nucleotides), added to junctions of rearranging immunoglobulin genes and T receptor genes in lymphocytes, may create premature termination codons when occurring within a functional reading frame (Li and Wilkinson, 1998). Cells of the immune system (and perhaps also other cell types) have evolved mechanisms to down-regulate mRNAs containing early termination codons. This regulatory response appears to be mediated by components of the translation machinery, in particular tRNAs and its operation are conditioned to a number of requirements (for example, introns must be found downstream of a nonsense codon). This mechanism has been coined "nonsense surveillance" (Li and Wilkinson, 1998) and it probably evolved to minimize perturbing effects of diversity-generation mechanisms needed to increase the immunoglobulin and T cell receptor repertoire in lymphocytes. Again, this recently unveiled biochemical mechanism clearly illustrates that large, complex genomes must avoid the lethal effects of increased mutagenesis to elude entry into error catastrophe (compare with Chapter 4).

Another occurrence of increased mutagenesis in differentiated organisms is cancer. In this case few restrictions seem to limit a cascade of mutagenesis events since processes of competition and selection promote survival and spread of the most fit transformed cells (Nicolson, 1987; Brash, 1997; Strauss, 1998). The genetic instability of transformed cells is the result of molecular alterations that dysregulate cell differentiation and cell division. Alterations in tumor suppressor genes, in DNA repair genes, microsatellite DNA modifications via expansion of trinucleotide and tetranucleotide repeats, oncogene activation and insertional mutagenesis, chromosome translocations, and abrogation of programmed cell death are among the types of events that may contribute to genetic instability and cell transformation (Broder, 1995; Naegeli, 1997; Jiricny, 1998). It is accepted that a number of genetic changes must occur in a normal cell to become a cancer cell. A number of human diseases in which there is a deffect in DNA damage recognition (e.g., *xeroderma pigmentosum*, Falconi's anemia, Bloom's syndrome, etc.) appear to be linked to cancer development (Naegeli, 1997). Normal diploid cells show a much higher genetic stability than cancer cells. Estimated mutation rates for mammalian cells are in the range of 10^{-10} substitutions per nucleotide site and cell generation (Loeb, 1994). This basal rate is increased in cancer cells, sometimes to the point of inhibiting the effectiveness of antitumor agents. Among the genes induced in transformed cells is the multidrug-transport protein P170 which exports drugs into the extracellular medium (Gottesman and Pastan, 1993). In agreement with currently accepted strategies of combination therapy for antiviral interventions, the advantages of combination anti-cancer therapy have long been recognized (Eckardt, 1985). The aim must be avoiding selection of drug-resistant cells by confronting them with multiple, simultaneous constraints (see Chapter 8), thus severely restricting the population size and the diversity of the pathogenic entities.

Evolution of Bacteria. Resistance to Multiple Antibiotics as a Long-Term, Undesigned Experiment

As surprising as it might seem, the belief that bacterial diseases were on their way to eradication was transmitted in classrooms world-wide during the mid 20th century. Even such an insightful mind as that of Sir Macfarlane Burnet wrote: "And since bacterial infections are, with unimportant exceptions, amenable to treatment with one or other of the new drugs, our real problems are likely to be concerned with virus diseases" (Burnet, 1966; page 360). The evolutionary potential for bacteria to become resistant to antibiotics was overlooked or minimized. And this occurred in spite of early evidence of selection of streptomycin-resistant mutants of *Mycobacterium tuberculosis* (Mitchison, 1950) and the realization that the association of several drugs given simultaneously prevented the emergence of resistant bacterial mutants (reviewed in O'Brien, 1993 and Heym et al, 1996). It has also been known for a long time that bacterial species are highly polymorphic (Milkman, 1973). However, genome complexity influences molecular evolutionary pathways employed to adapt to changing environments. Mutation rates of bacteria have been estimated to be about 10^{-9} substitutions per nucleotide (Drake, 1991), although subpopulations of pathogenic bacteria displaying increased mutation frequencies have been described and a number of mechanisms operate to increase mutation rates by several orders of magnitude in bacterial populations under stress associated with DNA lesions. In spite of a generally low mutation rate per nucleotide site, bacteria have exploited a number of molecular mechanisms to develop resistance to antibiotics, including horizontal gene transfers and chromosomal mutations (Chadwick and Goode, 1997; Mingeot-Leclerq et al, 1999). The latter mechanism operates with β-lactamase enzymes used by bacteria to inactivate β-lactam-containing antibiotics. Genes encoding β-lactamases are found in bacterial chromosomes, plasmids and transposons. Mutant β-lactamases have been selected that are capable of degrading different β-lactam types of antibiotics. A considerable number of residues in the enzyme molecule tolerate amino acid substitutions, and introduction of one mutation may trigger the fixation of others in order to compensate for loss of stability or catalytic activity (Huang and Palzkill, 1996; Zaccolo and Gherardi, 1999). In other cases, membrane proteins that mediate the import and export flux of small molecules including antibiotics in and out from the cell may be either altered or their level modulated to favor the export of the antibiotic.

Antibiotic resistance determinants may be encoded in plasmids, episomes, transposons or insertion sequences thus facilitating the transfer and spread of resistance genes (Chadwick and Goode, 1997; Scheld et al, 1998; Rice, 1998). One of the most dramatic examples is provided by the multiple resistance determinants found in *Staphylococcus aureus* strains and which are associated with an impressive spectrum of conjugative plasmids, episomes, transposons and insertion sequences. Letters of the English alphabet have been exhausted to name tetracycline resistance determinant(Levy et al, 1999).

Contrary to antiviral resistance, which generally affects only one viral lineage in its evolutionary adaptation to the inhibitor, the gene-transfer mechanisms in bacteria allow crossing of the species barriers. In this way, treatment of food products with antibiotics has prompted selection of nonpathogenic, antibiotic-resistant bacteria. The resistance determinants might then be transferred from the widespread, nonpathogenic bacteria into pathogenic bacteria. Although using different molecular vehicles and attaining overall different rates of implementation, there are clear parallelisms (also some differences; see Table 7.2) between the dominance of antiviral-resistant virus mutants and antibiotic-resistant bacteria. In both cases, in the face of an externally applied selective pressure, a number of molecular mechanisms come into play to select subpopulations of pathogens which are fit to replicate in the presence of the inhibitory constraint (Taylor and Feyereisen, 1996). Far from the predictions of M. Burnet (and many others), we have come to realize that the indiscriminate use of antibiotics has generated a new problem of intractable bacterial disease. This is currently aggravated by increasing numbers of immunocompromised individuals (mainly in underdeveloped countries) due to the expanding AIDS epidemics and also to malnutrition associated with social conflicts and proverty.

Evolutionary Potential of Unicellular Pathogens and Difficulties for the Control of Parasitic Disease

Drug-resistance is also becoming prominent among pathogenic, unicellular eukaryotes such as yeasts, amoebas, *Leishmania*, and the malaria parasite *Plasmodium falciparum*. Again, a variety of molecular mechanisms underlie generation of diversity and selection of inhibitor-resistant cells (Rex et al, 1995; Borst and Ouellette, 1995; Borst and Schinkel, 1997; Wahlgren et al, 1999). Point mutations in some genes that express proteins which are the target of drugs and activated ATP-dependent resistance efflux pumps are widespread mechanisms of multidrug resistance in eukaryotic cells. Not unexpectedly, unicellular parasites are highly polymorphic (Ben Miled et al, 1994; Gardner et al, 1996; Conway, 1997; Escalante et al, 1998). As in the

Table 7.2. A comparison of antibiotic resistance in bacteria and antiviral drug resistance in viruses

Distinctive features	
Antibiotics	Antiviral agents
• Maps on chromosomal genes, plasmids, mobile elements	• Maps on viral genome, often on target gene
• Cross-species transmission	• Often specific for virus genus
• Antibiotic use selects for resistance in nonpathogenic bacteria	• Selection of resistant forms confined to treated patients
• Resistance can be transferred from nonpathogenic to pathogenic bacteria	• Unlikely transfer of resistance to other viruses
Common features	
Antibiotics and Antiviral agents	

- Resistant forms gradually replace sensitive forms.
- Resistant pathogens often display decreased fitness relative to their sensitive counterparts, in the absence of the antibiotic or antiviral agent.
- Sensitive pathogens outcompete resistant ones in the absence of antibiotic or antiviral agent.
- Fitness decrease may be compensated by additional mutations in target genes.
- The degree of resistance may gradually increase in the continued presence of the antibiotic or antiviral agent.

case of viruses, growth conditions may perturb the equilibrium of natural parasite populations and promote the selective growth of subpopulations of the same parasite (Andrews et al, 1992). Exposed antigenic sites may be altered by point mutations and recombination events, resulting in resistance to monoclonal antibodies (Baltz et al, 1991; Al-Khedery et al, 1999). Polymorphisms and microdiversities represent a major problem for the design of antimalarial vaccines. There is evidence of positive selection for variation at the T cell epitopes located on relevant antigens of these parasites (De la Cruz et al, 1989). Growth of *Plasmodium falciparum* is inhibited by parasite-specific antibodies. Subpopulations of this parasite showing decreased sensitivity to monoclonal antibodies have been selected in vitro (Siripoon et al, 1997). The underlying mechanism could be either antibody-induced down-regulation of antigen expression, or selection of preexisting parasite subpopulations with constitutive, decreased expression of the relevant antigen. Induction of complement-resistance in some amoebae has been documented in vivo and in vitro (Hammelmann et al, 1993).

Drug-resistant variants of monocellular parasites have been selected by in vitro passage in the presence of sub-inhibitory concentrations of the drug (Peel et al, 1993, 1994; Lee et al, 1994; Berger et al, 1995; Bhasin and Nair, 1996). Selection of resistant parasite strains occur also upon drug treatments in vivo (Butcher et al, 1994). In analogy with current strategies for antiviral and antitumor strategies, the need to formulate antimalarial vaccines with mixtures of antigens reflecting multiple alleles of antigenic proteins has been recognized (Conway, 1997). It is not insignificant that insecticide-resistant mosquitoes are complicating problems with such insect-borne pathogens.

Although perhaps unexpected only a few decades ago, it is now obvious that RNA viruses and cells are endowed with the evolutionary potential to overcome selective constraints intended to limit their replication: The list of examples for parasites (even cancer cells can be regarded as parasites) given in this Chapter, although by no-means exhaustive, illustrates that they express their "virulence" with different rates, efficiencies, and mechanisms, but the result is a similar continuing threat to human health.

References

1. Agol VI. Recombination and other genomic rearrangements in picornaviruses. Seminars in Virology 1997; 8:77-84.
2. Agur Z, Kerszberg M. The emergence of phenotypic novelties through progressive genetic change. Am Nat 1987; 129:862-875.
3. Alcamí A, Symons JA, Khanna A et al. Poxviruses: Capturing cytokines and chemokines. Seminars in Virology 1998; 5:419-427.
4. Al-Khedery B, Barnwell JW, Galinski MR. Antigenic variation in malaria: a 3' genomic alteration associated with the expression of a *P. knowlesi* variant antigen. Cell 1999; 3:131-141.
5. Andersson DI, Hughes D. Muller's ratchet decreases fitness of a DNA-based microbe. Proc Natl Acad Sci USA 1996; 93:906-907.
6. Andrews RH, Chilton NB, Mayrhoefer G. Selection of specific genotypes of *Giardia intestinalis* by growth in vitro and in vivo. Parasitology 1993; 105:375-386.
7. Baltimore D. Retroviruses and retrotransposons; the role of reverse transcription in shaping the eukaryotic genome. Cell 1985; 40:481-482.
8. Baltz T, Giroud C, Bringand F et al. Exposed epitopes on a *Trypanosoma equiperdum* variant surface glycoprotein altered by point mutations. EMBO J 1991; 10:1653-1659.
9. Bebenek K, Abbotts J, Wilson SH et al. Error-prone polymerization by HIV-1 reverse transcripase. Contribution of template-primer misalignment, miscoding, and termination probability to mutational hot spots. J Biol Chem 1993; 268:10324-34.
10. Bell G. Sex and Death in Protozoa: The History of an Obsession. Cambridge: Cambridge University Press, 1988.
11. Ben Miled L, Dellagi K, Bernardi G et al. Genomic and phenotypic diversity of Tunisian *Theileria annulata* isolates. Parasitology 1994; 108:51-60.
12. Bennet A, Dao KM, Lenski R. Rapid evolution in response to high-temperature selection. Nature 1990; 346:79-81.
13. Berek C, Milstein C. The dymanic nature for the antibody repertoire. Immunol Rev 1988; 105:5-26.
14. Berger BJ, Carter NS, Fairlamb AH. Characterization of pentamidine-resistant *Trypanosoma brucei brucei*. Mol Biochem Parasitol 1995; 69:289-298.
15. Berkhout B, Klaver B, Das AT. Forced evolution of a regulatory RNA helix in the HIV-1 genome. Nucleic Acids Res 1997; 25:940-947.
16. Bhasin VK, Nair L. In vitro selection of *Plasmodium falciparum* lines resistant to dihydrofolate-reducatese inhibitors and cross resistance studies. Jpn J Med Sci Biol 1996; 49:1-14.
17. Borrego B, Novella IS, Giralt E et al. Distinct repertoire of antigenic variants of foot-and-mouth disease virus in the presence or absence of immune selection. J Virol 1993; 67:6071-6079.
18. Borrow P, Shaw GM. Cytotoxic T-lymphocyte escape viral variants: how important are they in viral evasion of immune clearance in vivo? Immunological Reviews 1998; 164:37-51.
19. Borst P, Ouellette M. New mechanisms of drug resistance in parasitic protozoa. Annu Rev Microbiol 1995; 49:427-460.
20. Borst P, Schinkel AH. Genetic dissection of the function of mammalian P-glycoproteins. Trends in Genet 1997; 13:217-222.
21. Brash DE. Sunlight and the onset of skin cancer. Trends in Genetics 1997; 13:410-414.
22. Broder S. Progress in the molecular medicine of cancer. In: Cooper GM, Temin RG, Sugden B, eds. The DNA Provirus. Howard Temin's Scientific Legacy. Washington DC: ASM Press 1995; 129-152.
23. Burnet M. Natural history of infectious disease. London: Cambridge University Press, 3rd edition, 1966.

24. Butcher PD, Cevallos AM, Carnaby S et al. Phenotypic and genotypic variation in *Giardia lamblia* isolates during chronic infection. Gut 1994; 35:51-54.
25. Carrillo C, Borca M, Moore DM et al. In vivo analysis of the stability and fitness of variants recovered from foot-and-mouth disease virus quasispecies. J Gen Virol 1998; 79:1699-1706.
26. Chadwick DJ, Goode J, eds. Antibiotic resistance: Origins, evolution, selection and spread. Ciba Foundation Symposium, Chichester: John Wiley and Sons Ltd., 1997.
27. Chan S-Y, Bernard H-U, Ong C-K et al. Phylogenetic analysis of 48 papillomavirus types and 28 subtypes and variants: A showcase for the molecular evolution of DNA viruses. J Virol 1992; 66:5714-5725.
28. Chao L. Fitness of RNA virus decreased by Muller's ratchet. Nature 1990; 348:454-55.
29. Clarke DK, Duarte EA, Moya A et al. Genetic bottlenecks and population passages cause profound fitness differences in RNA viruses. J Virol 1993; 67:222-228.
30. Clarke DK, Duarte EA, Elena S et al. The red queen reigns in the kingdom of RNA viruses. Proc Natl Acad Sci USA 1994; 91:4821-4824.
31. Conway DJ. Natural selection on polymorphic malaria antigens and the search for a vaccine. Parasitology Today 1997; 13:26-29.
32. De la Cruz VF, Maloy WL, Miller LH et al. The immunologic significance of variation within malaria circumsporozoite protein sequences. J Immunol 1989; 142:3568-3575.
33. Domingo E. Biological implications of the quasispecies populations of RNA viruses. Proceedings of the VIII International Symposium on viral hepatitis. Hepatología Clínica 6, suppl. 1998; 1:59-64.
34. Domingo E, Sabo DL, Taniguchi T et al. Nucleotide sequence heterogeneity of an RNA phage population. Cell 1978; 13:735-744.
35. Domingo E, Mateu MG, Martínez MA et al. Genetic variability and antigenic diversity of foot-and-mouth disease virus. Applied Virology Research 1990; 2:233-266.
36. Domingo E, Escarmís C, Martínez MA et al. Foot-and-mouth disease virus populations are quasispecies. Current Topics in Microbiology and Immunology 1992; 176:33-47.
37. Domingo E, Holland JJ. Mutation rates and rapid evolution of RNA viruses. In: Morse SS, ed. Evolutionary Biology of viruses. New York: Raven Press 1994:161-184.
38. Domingo E, Escarmís C, Sevilla N et al. Basic concepts in RNA virus evolution. FASEB J 1996; 10:859-864.
39. Domingo E, Holland JJ. RNA virus mutations and fitness for survival. Annu. Rev. Microbiol 1997; 51, 151-78.
40. Domingo E, Baranowski E, Ruiz-Jarabo CM et al. Quasispecies structure and persistence of RNA viruses. Emerging Infectious Diseases 1998; 4:521-527.
41. Drake JW. A constant rate of spontaneous mutation in DNA-based microbes. Proc Natl Acad Sci USA 1991; 88:7169-7164.
42. Drake JW. Rates of spontaneous mutations among RNA viruses. Proc Natl Acad Sci USA 1993; 90:4171-4175.
43. Duarte E, Clarke D, Moya A et al. Rapid fitness losses in mammalian RNA virus clones due to Muller's ratchet. Proc Natl Acad Sci USA 1992; 89:6015-19.
44. Duarte EA, Clarke DK, Moya A et al. Many trillionfold amplification of single RNA virus particles fails to overcome the Muller's ratchet effect. J Virol 1993; 67:3620-623.
45. Eckhardt S. Potentiation of antimetabolite action by alkylating agents. Adv in Enzyme Regul 1985; 24:143-153.
46. Elena SF, Dávila M, Novella IS et al. Evolutionary dynamics of fitness recovery from the debilitating effects of Muller's ratchet. Evolution 1998; 52; 309-314.
47. Erwin DH, Valentine JW. "Hopeful monsters", transposons and Metazoan radiation. Proc Natl Acad Sci USA 1984; 81:5482-5483.
48. Escalante AE, Lal AA, Ayala FJ. Genetic polymorphism and natural selection in the malaria parasite *Plasmodium* falciparum. Genetics 1998; 149:189-202.
49. Escarmís C, Dávila M, Charpentier N et al. Genetic lesions associated with Muller's ratchet in an RNA virus. J Mol Biol 1996; 264:255-67.
50. Escarmís C, Dávila M, Domingo E. Multiple molecular pathways for fitness recovery of an RNA virus debilitated by operation of Muller's ratchet. J Mol Biol 1999; 285:495-505.
51. Fisher RA. The genetical theory of natural selection. Oxford: Oxford University Press, 1930.

52. Gabriel A, Willems M, Mules EH et al. Replication infidelity during a single cycle of Ty 1 retrotransposition. Proc Natl Acad Sci USA 1996; 93:7767-7771.
53. Garcia-Maruniak A, Pavan OHO, Maruniak JE. A variable region of *Anticarsia gemmatalis* nuclear polyhedrosis virus containd tandemly repeated DNA sequences. Virus Res 1996; 41:123-132.
54. Gardner JP, Pinches RA, Roberts DJ et al. Variant antigens and endothelial receptor adhesion in *Plasmodium falciparum*. Proc Natl Acad Sci USA 1996; 93:3503-3508.
55. Gause GF. The Struggle for Existence. New York: Dover Publications Inc. 1971.
56. Gottesman MM, Pastan I. Biochemistry of multidrug resistance mediated by the multidrug transporter. Annu Rev Biochem 1993; 62:385-427.
57. Goudsmit J, de Ronde D, Ho DD et al. Human immunodeficiency virus fitness in vivo: Calculations based on a single zidovudine resistance mutation at codon 215 of reverse transcriptase. J Virol 1996; 70:5662-5666.
58. Grebogi C, Hammel SM, Yorke JA et al. Shadowing of physical trajectories in chaotic dynamics: containment and refinement. Phys Rev Letters 1990; 65:1527-1530.
59. Gromeier M, Wimmer E, Gorbalenya AE. Genetics, pathogenesis and evolution of picornaviruses. In: Dominge E, Webster RG, Holland JJ, eds. Origin and Evolution and Viruses. San Diego: Academic Press 1999; 287-343.
60. Hamelmann C, Foerster B, Burchard GD et al. Induction of complement resistance in cloned pathogenic *Entamoeba histolytica*. Parasite Immunol 1993; 15:223-228.
61. Harrigan PR, Bloor S, Larder BA. Relative replicative fitness of zidovudine-resistant human immunodeficiency virus type 1 isolates in vitro. J Virol 1998; 72:3773-3778.
62. Hill AVS. The immunogenetics of human infectious diseases. Annu Rev Immunol 1998; 16:593-617.
63. Heym B, Philipp W, Cole ST. Mechanisms of drug resistance in *Mycobacterium tuberculosis*. Curr Top Microbiol Immunol 1996; 215:49-69.
64. Holland JJ, de la Torre JC, Steinhauer DA et al. Virus mutation frequencies can be greatly underestimated by monoclonal antibody neutralization of virions. J Virol 1989; 63, 5030-36.
65. Holland JJ, de la Torre JC, Clarke DK et al. Quantitation of relative fitness and great adaptability of clonal populations of RNA viruses. J Virol 1991; 65:2960-67.
66. Holland JJ, de la Torre JC, Steinhauer DA. RNA virus populations as quasispecies. Curr Top Microbiol Immunol 1992; 176:1-20.
67. Holland JJ, Domingo E. Origin and evolution of viruses. Virus Genes 1998; 16:13-21.
68. Huang W, Palzkill T. Amino acid sequence determinants of β-lactamase structure and activity. J Mol Biol 1996; 258:688-703.
69. Isnard M, Granier M, Frutos R et al. Quasispecies nature of three maize streak virus isolates obtained through different modes of selection from a population used to assess response to infection of maize cultivars. J Gen Virol 1998; 79:3091-3099.
70. Jiricny J. Replication errors: cha(lle)nging the genome. The EMBO J 1998; 17:6427-6438.
71. Kallberg E, Jainanduninsing S, Gray D, Leanderson T. Somatic mutation of immunoglobulin V genes in vitro. Science 1996; 271:1285-1288.
72. Kimata JT, Kuller R, Anderson DB69. Emerging cytopathic and antigenic simian immunodeficiency virus variant influence AIDS progression. Nature Medicine 1999; 5:535-541.
73. Kimura M. The Neutral Theory of Molecular Evolution. Cambridge: Cambridge University Press, 1983.
74. Kunkel TA. Misalignment-mediated DNA synthesis errors. Biochemistry 1990; 29:8003-8011.
75. Lednicky JA, Arrington AS, Stewart AR et al. Natural isolates of simian virus 40 from immunocompromised monkeys display extensive genetic heterogeneity: New implications for polyomavirus disease. J Virol 1998; 72:3980-3990.
76. Lee ST, Lin HY, Lee SP et al. Selection of arsenite resistance causes reversible changes in minicircle composition and kinetoplast organization in *Leishmania mexicana*. Mol Cell Biol 1994; 14:587-596.
77. Lenski R, Travisano M. Dynamics of adaptation and diversification: A 10,000-generation experiment with bacterial populations. Proc Natl Acad Sci USA 1994; 91:6808-6814.
78. Leslie JF, Vrijenhoek RC. Consideration of Muller's ratchet mechanism through studies of genetic linkage and genomic compatibilies in clonally reproducing *Poeciliopsis*. Evolution 1980; 34:1105-15.
79. Levy SB, McMurry LM, Barbosa TM et al. Nomenclature for new tetracycline resistance determinants. Antimicrob Agents Chemother 1999; 43:1523-1524.

80. Li S, Wilkinson MF. Nonsense surveillance in lymphocytes? Immunity 1998; 8:135-141.
81. Licis N, van Duin J, Blaklava Z, Berzins V. Long-range translational coupling in single-stranded RNA bacteriophages: an evolutionary analysis. Nucleic Acids Res 1998; 26:3242-3246.
82. Loeb LA. Microsatellite instability: Marker of a mutator phenotype in cancer. Cancer Res 1994; 54:5059-5063.
83. Londsdale DM, Brown SM, Lang J et al. Variations in herpes simplex virus isolated from human ganglia and a study of clonal variation in HSV-1. Ann NY Acad Sci 1980; 354:291-308.
84. Martínez MA, Carrillo C, González-Candelas F et al. Fitness alteration of foot-and-mouth disease virus mutants: measurement of adaptability of viral quasispecies. J Virol 1991; 65:3954-3957.
85. Matsuda T, Bebenek K, Masutani C et al. Low fidelity DNA synthesis by human DNA polymerase-η. Nature 2000; 404:1011-13.
86. McClure MA. The retroid agents: Disease, function and evolution. In: Domingo E, Webster RG, Holland JJ, eds. Origin and Evolution of Viruses. London: Academic Press, 1999; in press.
87. McGeoch DJ, Davison AJ. The molecular evolutionary history of the herpesviruses. In: Domingo E, Webster RG, Holland JJ, eds. Origin and Evolution of Viruses. San Diego: Academic Press, 1999:441-465.
88. Meyerhans A, Vartanian J-P. The fidelity of cellular and viral polymerases and its manipulation for hypermutagenesis. In: Domingo E, Webster RG, Holland JJ, eds. Origin and Evolution of Viruses. San Diego: Academic Press 1999: 87-114.
89. Milkman R. Electrophoretic variation in *Escherichia coli* from natural sources. Science 1973; 182:1024-1026.
90. Mingeot-Leclercq M-P, Glupczynski Y, Tulkens PM. Aminoglycosides: activity and resistance. Antimicrob Agents and Chemother 1999; 43:727-737.
91. Mitchison DA. Development of streptomycin resistant strains of tubercle bacilli in pulmonary tuberculosis. Thorax 1950; 5:144-61.
92. Muller HJ. The relation of recombination to mutational advance. Mutat Res 1964; 1:2-9.
93. Muñoz D, Murillo R, Knell PJ et al. Four genotypic variants of *Spodoptera exigua Nucleopolyhedrovirus* (Se-SP2) are distinguishable by a hypervariable genomic region. Virus Res 1999; 59:61-74.
94. Murphy PM. Molecular mimicry and the generation of host defense protein diversity. Cell 1993; 72:823-826.
95. Naegeli H. Mechanisms of DNA Damage Recognition in Mammalian Cells. Austin: RG Landes Co; 1997.
96. Nicolson GL. Tumor cell instability, diversification, and progression to the metastatic phenotype: from oncogene to oncofetal expression. Cancer Res 1987; 47:1473-1487.
97. Nijhuis M, Schuurman R, de Jong D et al. Increased fitness of drug resistant HIV-1 protease as a result of acquisition of compensatory mutations during suboptimal therapy. AIDS 1999; 13:2349-59.
98. Novella IS, Duarte EA, Elena SF et al. Exponential increases of RNA virus fitness during large population transmissions. Proc Natl Acad Sci USA 1995b; 92:5841-44.
99. Novella IS, Elena SF, Moya A et al. Size of genetic bottlenecks leading to virus fitness loss is determined by mean initial population fitness. J Virol 1995b; 69:2869-72.
100. Novella IS, Hershey CL, Escarmís C et al. Lack of evolutionary stasis during alternating replication of an arbovirus in insect and mammalian cells. J Mol Biol 1999a; 287:459-465.
101. Novella IS, Quer J, Domingo E et al. Exponential fitness gains of RNA virus populations are limited by bottleneck effects. J Virol 1999b; 73:1668-71.
102. O'Brien RJ. The treatment of tuberculosis. In: Reichman LB, Hershfield ES, eds. Tuberculosis—a comprehensive international approach. New York: Dekker 1993; 207-240.
103. Page RDM, Holmes EC. Molecular Evolution. A phylogenetic approach. Oxford: Blackwell Science Ltd, 1998.
104. Parrish CR, Truyen U. Parvoviruses variation and evolution. In: Domingo E, Webster RG, Holland JJ, eds. Origins and Evolution of Viruses. San Diego: Academic Press, 1999:421-39.
105. Peel SA, Merrit SC, Handy J et al. Derivation of highly mefloquine-resistant lines from *Plasmodium falciparum* in vitro. Am J Trop Med Hyg 1993; 48:385-397.

106. Peel SA, Bright P, Yount B et al. A strong association between mefloquine and halofantrine resistance and amplification, overexpression, and mutation in the P-glycoprotein gene homolog (*pfmdr*) of *Plasmodium falciparum* in vitro. Am J Trop Med Hyg 1994; 51:648-658.
107. Quer J, Huerta R, Novella IS et al. Reproducible nonlinear population dynamics and critical points during replicative competitions of RNA virus quasispecies. J Mol Biol 1996; 264:465-471.
108. Rex JH, Rinaldi MG, Pfaller MA. Resistance of *Candida* species to fluconazole. Antimicrob Agents Chemother 1995; 39:1-8.
109. Reznick D, Travis J. The empirical study of adaptation in natural populations. In: Rose MR, Lauder GV, eds. Adaptation. San Diego: Academic Press 1996; 243-89.
110. Rice LB. Tn*916* family conjugative transposons and dissemination of antimicrobial resistance determinants. Antimicrob Agents and Chemother 1998; 285:775-783.
111. Rojas JM, Dopazo J, Martín-Blanco E et al. Analysis of genetic variability of populations of herpes simplex viruses. Virus Res 1993; 28:249-261.
112. Rojas JM, Dopazo J, Nájera I et al. Molecular epidemiology of HIV-1 in Madrid. Virus Res 1994; 31:331-342.
113. Ruiz-Jarabo CM, Sevilla N, Dávila M et al. Antigenic properties and population stability of a foot-and-mouth disease virus with an altered Arg-Gly-Asp receptor-recognition motif. J Gen Virol 1999; 80:1899-1909.
114. Sakaoka H, Kurita K, Gouro T et al. Analysis of genomic polymorphism among herpes simplex virus type 2 isolates from four areas of Japan and three other countries. J Med Virol 1995; 45:259-272.
115. Sarisky RT, Nguyen TT, Duffy KE et al. Difference in incidence of spontaneous mutations between herpes simplex virus types 1 and 2. Antimicrob Agents Chemother 2000; 44:1524-29.
116. Scheld JM, Armstrong D, Hughes JM, eds. Emerging Infections 1. Washington DC: American Society for Microbiology Press 1998.
117. Scott TW, Weaver SC, Mallampali VL. Evolution of mosquito-borne viruses. In: Morse SS, ed. Evolutionary Biology of Viruses. New York: Raven Press 1994; 293-324.
118. Sevilla N, Ruiz-Jarabo CM, Gómez-Mariano G et al. An RNA virus can adapt to the multiplicity of infection. J Gen Virol 1998; 79:2971-2980.
119. Siripoon IJ, Snounou VG, Perlmann P et al. Plasmodium falciparum: selection of parasite subpopulations with decreased sensitivity for antibody-mediated growth inhibition in vitro. Parasitology 1997; 114:317-324.
120. Smith DB, Inglis SC. The mutation rate and variability of eukaryotic viruses: an analytical review. J Gen Virol 1987; 68:2729-2740.
121. Smith GL, Symons JA, Alcamí A. Poxviruses: Interfering with interferon. Seminars in Virology 1998; 8:409-418.
122. Storb U. The molecular basis of somatic hypermutation of immunoglobulin genes. Curr Op Immunol 1996; 8:206-214.
123. Strauss BS. Hypermutability in carcinogenesis. Genetics 1998; 148:1619-1626.
124. Sugimoto C, Kitamura T, Guo J et al. Typing of urinary JC virus DNA offers a novel means of tracing human migrations. Proc Natl Acad Sci USA 1997; 94:9191-9196.
125. Taylor M, Feyereisen R. Molecular biology of resistance to toxicans. Mol Biol Evol 1996; 13:719-734.
126. Umene K, Watsom RJ, Enquist LW. Tandem repeat DNA in an intergenic region of herpes simplex virus type 1 (Patton). Gene 1984; 30:33-39.
127. Valcarcel J, Ortin J. Phenotypic hiding: the carry-over of mutations in RNA viruses as shown by detection of *mar* mutants in influenza virus. J Virol 1989; 63:4107-09.
128. Van Ranst M, Kaplan JB, Sundbery JP et al. Molecular evolution of papillomaviruses. In: Gibbs A, Calisher CH, Garcia-Arenal F, eds. Molecular Basis of Virus Evolution. Cambridge: Cambridge University Press 1995; 455-476.
129. van Valen L. A new evolutionary law. Evol Theory 1973; 1:1-30.
130. Villarreal LP. DNA virus contribution to host evolution. In: Domingo E, Webster RG, Holland JJ, eds. Origin and Evolution of Viruses. San Diego: Academic Press, 1999: 391-420.
131. Wahlgren M, Fernández V, Chen Q et al. Waves of malarial *var*-iations. Cell 1999; 96:603-606.

132. Weaver SC. Recurrent emergence of Venezuelan equine encephalomyelitis. In: Sheld WM, Hughes J, eds. Emerging Infections I. Washington DC: American Society for Microbiol Press, 1998; 27-42.
133. Weaver SC, Rico-Hess, R, Scott TW. Genetic diversity and slow rates of evolution in new world alphaviruses. Curr Top Microbiol Immunol 1992; 176:99-117.
134. Weaver SC, Brault AC, Kang W et al. Genetic and fitness changes accompanying adaptation of an arbovirus to vertebrate and invertebrate cells. J Virol 1999; 73:4316-4326.
135. Wells, RD, Bacolla A, Bowater RP. Instabilities of triplet repeats: Factors and mechanisms. In: Oostra BA, ed. Trinucleotide diseases and instability. Berlin: Springer 1998; 133-165.
136. Williams GC. Natural Selection. Domains, Levels, and Challenges. New York, Oxford: Oxford University Press, 1992.
137. Wimmer E, Hellen CUT, Cao X. Genetics of poliovirus. Annu Rev Genet 1993; 27:353-436.
138. Winocour E, Keshet I, Nedjar G et al. Origins of SV40 genetic variation. In: Palese P, Roizman B, eds. Genetic Variation of Viruses. New York: The New York Academy of Sciences 1980; 43-52.
139. Woodbury EL, Samuel R, Knowles NJ. Serial passages in tissue culture of mixed foot-and-mouth disease virus serotypes. Arch Virol 1995; 140:783-787.
140. Wright S. Evolution in Mendelian populations. Genetics 1931; 16:97-159.
141. Wright S. Character change, speciation, and the higher taxa. Evolution 1982; 36:427-43.
142. Yamaguchi T, Yamashita Y, Kasamo K et al. Genomic heterogeneity maps to tandem repeat sequences in the herpes simplex virus type 2 U_L region. Virus Res 1998; 55:221-231.
143. Yáñez R, Moya A, Viñuela E, Domingo E. Repetitive nucleotide sequencing of a dispensable DNA segment in a clonal population of African swine fever virus. Virus Res 1991; 20, 265-272.
144. Yuste E, Sánchez-Palomino S, Casado C et al. Drastic fitness loss in human immunodeficiency virus type 1 upon serial bottleneck events. J Virol 1999; 73:2745-2751.
145. Zaccolo M, Gherardi E. The effect of high-frequency random mutagenesis on in vitro protein evolution: A study on TEM-1 β-lactamase. J Mol Biol 1999; 285:775-783.
146. Zennou V, Mammano F, Paulous S et al. Loss of viral fitness associated with multiple Gag and Gag-Pol processing defects in human immunodeficiency virus type 1 variants selected for resistance to protease inhibitors in vivo. J Virol 1998; 72:3300-3306.
147. Zong J-C, Ciufo DM, Alcendor DJ et al. High-level variability in the ORF-K1 membrane protein gene at the left end of the Kaposi's sarcoma-associated herpesvirus genome defines four major virus subtypes and multiple variants or clades in different human populations. J Virol 1999; 73:4156-4170.

CHAPTER 8

Connections, Implications and Prospects
Virus Transmission and Divergence in Natural Infections: Tolerances and Constraints

Viruses undergo genetic change in each infected individual, pushed by mutational pressure and guided by the interplay between positive and negative selection, as discussed in preceding Chapters. The next step into the process of long-term evolution of viruses is transmission from an infected host into a susceptible one. Transmission may exert an important influence on viral evolution since it may occur between two individuals of the same host species, horizontally or vertically, or quite disparate species as with arboviruses. In the alphavirus Venezuelan equine encephalitis virus (VEEV) viral titers in the blood of horses reach 10^8 infectious units per ml, or a total of about 3×10^{12} infectious units in the blood of a single animal. These high titers probably facilitate transmission to insect vectors by uptake of blood. In humans, virus levels in blood are much lower, and therefore humans are dead-end hosts for VEEV infections (Weaver, 1998) (Fig. 8.1). Transmission may involve massive amounts of virus (as in insects which have taken blood meals recently from viremic animals, or in human transfusions with contaminated blood) or one or few particles, such as in aerosol transmission, or from mosquitoes long after they have taken blood meals. In the latter case, transmission constitutes a natural bottleneck. A number of models have been developed to describe the spread of viral pathogens in interaction with their hosts (Anderson and May, 1991; May 1993, 1995; Garnett and Antia, 1994; Ewald, 1994; Moya and García-Arenal, 1995; Kaslow and Evans, 1997). Recently, the evolution of viral virulence upon horizontal versus vertical transmission has been modeled with regard to the quasispecies dynamics of the infectious agent (Bergstrom et al, 1999), and the predictions of this model are in good agreement with experimental observations on fitness variations as reviewed in Chapter 7.

Consideration of the pathogenic agent not as a defined entity in which genomes monotonously repeat the same information, but as a distribution of nonidentical genomes with potentially different phenotypes (the quasispecies structure described in previous Chapters) renders long-term evolution a complex and quite unpredictable process. A transmission bottleneck will isolate a single infectious genome in a new host, constituting a new start of an evolutionary process which may be influenced by the nature of the founder genome. The initial position in sequence space (in this case represented by one individual genome initiating viral multiplication following transmission) is one of the parameters which may have an effect on virus adaptation (Domingo et al, 1999). When a massive amount of virus is transmitted (such as in transfusion of viremic blood), a competition among many infectious genomes may be established, with all its implications for rapid adaptation of the virus in the new individual. The varia-

---

Quasispecies and RNA Virus Evolution: Principles and Consequences, edited by Esteban Domingo, Christof K. Biebricher, Manfred Eigen and John J. Holland. ©2001 Eurekah.com.

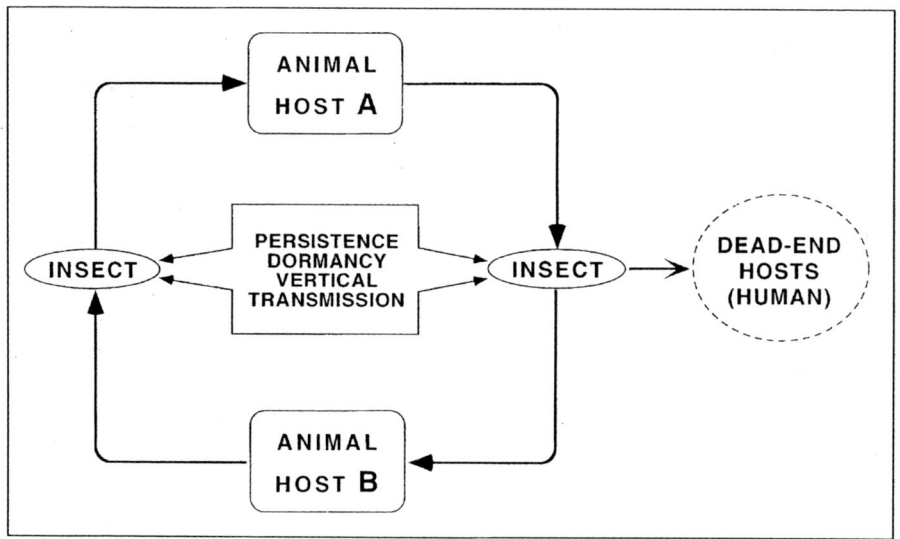

Fig. 8.1. Complexity of the arbovirus life cycle in which viruses alternate between insect hosts (where they may persist for prolonged time periods) and animal hosts (where they produce a systemic infection). Examples are the alphaviruses for which humans are dead-end hosts because viremia (virus level in blood) is insufficient for viruses to be uptaken and transmitted by insects.

tions and indeterminacy involved in viral transmission can be illustrated with a number of quantitative considerations. Aerosol transmission of viruses (such as the common cold or influenza viruses) mostly involves droplets of a size in the range of 0.1 to 10 µm in diameter (Artenstein and Miller, 1996; Gerone et al, 1966). In measurements with adult human volunteers, the titers of influenza virus in nasopharyngeal washings peaked at 48 h post-infection and titers were in the range of 10^3-10^8 tissue culture infectious doses per ml (Murphy and Webster, 1996). Therefore in the event of aerosol transmission, most droplets would contain either no virus or one infectious particle, and less frequently, several infectious particles. Blood from patients infected with hepatitis B virus (HBV), hepatitis C virus (HCV) or human immunodeficiency virus (HIV) may contain a few or up to hundreds of thousands infectious units per milliliter, and lower amounts of virus may permeate some body fluids. Therefore the types of contacts that lead to exposure to (and transmission of) a virus produce infecting viral doses that can range over many orders of magnitude. Since the complexity of the viral quasispecies in the infecting host may also vary over at least two orders of magnitude [depending of the type of virus, distance from a clonal origin and history of the infection (acute, chronic, etc.) (Domingo and Holland, 1997)] the population basis for the unpredictability of the course of individual infections rests upon clear quantitative grounds (Holland et al, 1992).

Functional and structural constraints limit the diversification of viruses in nature, as they expand over decades into new individual hosts in successive outbreaks, epidemics and recurrent epidemics. Restrictions can be observed at three levels: the viral genotype, encoded proteins and phenotypic traits. Examples at the level of the viral genotype are biases in certain types of mutations (synonymous versus nonsynonymous, transitions versus transversions, absence of additions or deletions, etc.) including preference for some bases at the third position of codons due to biased codon usage by the translation apparatus. At the protein level, restrictions are often evidenced by a limited repertoire of the amino acids found even at highly variable sites of viral proteins. At the phenotypic level, some viruses can readily adapt to a given modified environment while others are absolutely unable to adapt. There are numerous examples in the literature illustrating these three types of restrictions, and on occasions the functional basis for the restriction has been established. The need to maintain a range of secondary and tertiary structures in the IRES element in picornaviruses (Chapter 3) forces the occurrence of compensatory mutations and limits greatly the possibilities of genetic drift of this regulatory element. Manipulation of RNA phage genomes has often indicated that folding patterns of RNA, again necessitating specific base-base interactions, are needed for infectivity (Arora et al, 1996). Folding restrictions at the RNA level may result in limitation of the number of mutations at silent sites of codons within open-reading frames (Domingo, 1992).

Regarding viral proteins, even a highly variable domain such as the V3 of the surface glycoprotein gp120 HIV-1 shows abundance of some types of amino acid replacements which are not the result of biases derived from codon usage (Lamers et al, 1996; Penny et al, 1996; Korber et al, 1998). In a study of the evolution of FMDV over a period of six decades, it was observed that while an irregular (not constant with time) accumulation of synonymous mutations had taken place, there was no net accumulation of amino acid replacements (Martínez et al, 1992). For many cellular and viral genes the rate of fixation of synonymous mutations is three- to six-fold higher than for nonsynonymous replacements (Nei, 1987). However, in FMDV a more subtle picture of long-term evolution has emerged. In the survey of 30 isolates obtained over six decades, some variable positions within two major antigenic sites showed alternation of amino acids of the type: amino acid 1 (first decade) → amino acid 2 (intermediated decades) → amino acid 1 (recent decade) (Martínez et al, 1991, 1992). The main conclusions of these analyses are that antigenic variation can occur without a linear accumulation of amino acid replacements, and that FMDV has exploited in its natural evolution only a minimum of its potential for antigenic variation.

There are some general features of RNA viruses that help in explaining the origins of some restrictions that limit their evolution. RNA viruses have compact genomes and the same nucleotide sequences may serve a range of structural and functional roles. Designs to achieve maximum compactness include overlapping reading frames, frameshifting, alternative splicing, RNA editing, ambisense RNA, utilization of host-encoded proteins for replication, etc. Furthermore, some viral proteins play multiple roles in the viral life cycle. Perhaps one of the most spectacular examples of a multifunctional protein is provided by the coat protein of some plant viruses of the *Bromoviridae* family. For these viruses, the coat protein is involved in initiation of infection via binding to a stem-loop structure near the 3' end (and perhaps also to internal sites) of genomic RNAs, release of plus-strand RNA from replication complexes, subgenomic RNA synthesis, and cell to cell movement within plants, in addition to its canonical role in virion assembly and as a structural protein (Bol, 1999).

Constraints also have a historical component since the evolution of a virus (as that of any biological entity) is dependent on the modular context under which its replicative functions have been shaped. These historical constraints are reminiscent of the developmental constraints that must influence the evolution of differentiated organisms (Maynard Smith et al, 1985). Restrictions acting at the RNA and protein level, limit the occupation of sequence space (Domingo and Holland 1997). Critical issues remaining to be investigated are to define the molecular and functional basis for such restrictions, and also to establish under which circumstances such restrictions can be released thereby expanding occupation of sequence space by a virus. This latter point is of obvious relevance to the emergence of new viral pathogens, to be discussed in a later section of this Chapter. A highly conserved Arg-Gly-Asp (RGD) triplet at one of the surface loops of FMDV plays a dual role: it serves to recognize integrins, one of the receptor types utilized by FMDV for entry into the cell (Chapter 2), and it provides critical residues for the binding of several neutralizing antibodies (Verdaguer et al 1995, 1996, 1998). This dual involvement came as a surprise since it had been widely accepted that receptor-recognition sites in picornaviruses must be hidden from immune attack. In spite of being part of several relevant epitopes, the conservation of the RGD triplet was imposed by the need to recognize a receptor, an essential event for viral infection. Yet in an FMDV which was multiply passaged in cell culture, the virus acquired the ability to use an alternative receptor thereby rendering the RGD dispensable (Martínez et al, 1997). The basis for this critical shift was the fixation of a few amino acid replacements on the viral capsid. This illustrates how an evolving quasispecies, by virtue of unceasing mutational pressure, and tolerance for specific capsid mutations, can trigger expansions in the occupation of sequence space by variations at previously invariant domains. At present, mutants of FMDV including RGG or even GGG instead of RGD have been identified, and they show profound alterations in cell tropism (Baranowski et al, 1998; Ruiz-Jarabo et al, 1999). Changes of cell tropism associated with a few amino acid substitutions have been observed with many viruses, and the role that such shifts might have in viral pathogenesis, and in the emergence of new viral pathogens, constitutes an active field of research.

Had individual, geographically independent isolates of the same virus genus been sequenced while ignoring quasispecies as portrayed throughout this book, sequence variations would have been rightly equated with the concept of genetic polymorphism as presented in classical population genetics. It is important to attempt to define conceptual limits to quasispecies versus polymorphisms since this may help in clarifying the novelty and practical problems inherent to the quasispecies organization of RNA viruses.

Relationships of Quasispecies with Population Genetics and Current Concepts of Complexity

Population Genetics

Genetic polymorphisms were discovered in the 1960's by Lewontin and Hubby working with *Drosophila* and by Harris with humans (Lewontin and Hubby, 1996; Harris, 1966; overviews in Lewontin, 1974; Ayala, 1976). These early studies employed the zymogram technique which is based upon the electrophoretic separation of different forms of the same enzyme followed by in situ enzyme activity determinations. Different individuals of the same animal or plant species occasionally harbored a variant form of an enzyme, and although effects of post-translational modifications were generally not ruled out, it was correctly inferred that many loci contained alternative forms (alleles) of the same gene. These observations represented a turning point in population biology since the level of polymorphism, or the number of different forms of the same gene in populations of differentiated organisms, proved to be much higher than had been anticipated. Later, an even higher level of genetic heterogeneity was revealed by application of gene cloning and rapid nucleotide sequencing techniques. Most differentiated organisms are polymorphic at many sites, individual humans being distinguished on average, in one out of 200 to 400 nucleotides in their chromosomal DNA. Although most of these variations have little or no effect, and go unnoticed, some of them are associated with genetic disorders or propensities to some diseases (Cooper and Krawczak, 1993; Cargill et al, 1999). Polymorphisms in genes related to immune responses are one of the major influences on the different susceptibilities of individual hosts of the same animal species when infected by the same pathogen (Chapter 6).

A conceptual difference between quasispecies as applied to genetically simple replicons and polymorphisms as applied to cellular organisms is implied in the original definitions of the two terms, and also in the different biological implications when analyzed in quantitative terms. It is noteworthy that in classical population studies, alleles that were found only once were not counted as entering into the definition of polymorphism (see, for example, Spiess, 1977, page 90). Any sampling of a mutant distribution in a viral quasispecies reveals

Table 8.1. Parameters relevant to the adaptive potential of RNA viral quasispecies[a]

- GENETIC HETEROGENEITY OF RNA GENOME POPULATIONS
Variable, depending on virus and evolutionary history. Generally it is in the range of 1 to 100 mutations per genome.
- VIRUS POPULATION SIZE
Variable. In acutely infected organisms infectious particles may reach 10^9 to 10^{12} at any given time. In a viral plaque on a cell monolayer infectious units are 10^3 up to 10^{10}, depending on the virus and host cell and plaque size.
- GENOME LENGTH
3 Kb to 30 Kb
- MUTATIONS NEEDED FOR RELEVANT PHENOTYPIC MODIFICATIONS
For a number of adaptive changes one or very few nucleotide or amino acid replacements are sufficient.

[a]Based in data reviewed in Domingo et al, 1985 and Domingo, 1996

many rare (unique) mutants due merely to mutational pressure, irrespective of their selective value. Rare, unique forms of a viral genome that constitute integral parts of mutant distributions would be excluded if one adhered to the classical definition of polymorphism. Some virologists working on HIV are currently establishing a difference between "polymorphic" sites and the microheterogeneities associated with the quasispecies structure. Although this distinction may at times be fuzzy, it is a correct and useful application of the original concept of polymorphism. Yet, the mutant spectrum, a hallmark of quasispecies structure and dynamics of RNA viruses, was not considered in the original definition of polymorphism. Therefore to dispute the term quasispecies on the grounds that it is phenomenologically similar to classical polymorphisms is a gross oversimplification which fails to understand the dynamic functional role of shifting mutant spectra in large populations of RNA viruses (Domingo et al, 1995).

A second, perhaps more relevant, difference between quasispecies and polymorphism, refers to their widely disparate quantitative impact for the populations of replicons to which the two concepts commonly apply (see Fig. 5.5 in Chapter 5). The quantitative impact of a mutant distribution in a viral quasispecies can only be assessed by considering four related parameters (Table 8.1). It is the relationship between the genome length and average mutation frequencies that renders simple replicons unique as compared to cells. A defined viral genome of 10,000 nucleotides has a total of 3×10^4 possible single mutants, even though many of them will never survive due to low fitness or lethality. This potential number of single mutants is below the size of many viral populations: even a single plaque of foot-and-mouth disease virus or vesicular stomatitis virus on a cell monolayer contains 10^5 and 10^{10} infectious units, respectively, and infected animals may contain in excess of 10^{12} infectious units. For a mammalian genome the potential number of single mutants is in the range of 10^{10}, which is well above the population size of mammalian species. Therefore, the capacity to explore sequence space is vastly more extensive for a simple replicon with the high mutation rates and levels of heterogeneity seen in RNA viruses. All share structural and functional constraints that require them to be defined clouds of sequence space. Yet the parameters compared in Table 8.1 render RNA genetic elements highly dynamic and adaptable with regard to a relatively (only relatively) more static DNA world (Holland et al, 1982).

Differences between the theoretical concept of quasispecies as originally formulated, and its actual application to describe virus populations, has at times been considered a weakness in the use of quasispecies by experimental virologists. In this respect, it should not be forgotten that many such important concepts (in physics, chemistry and biology) have been modified and adapted to find new, interesting applications. To mention an example of population biology, no real population fulfills the requirement of being ideal, closed and panmictic (in which individuals mate at random). Yet population biologists analyze many natural populations and interpret results on the basis of such simplifying assumptions. Also, the increasingly influential metapopulation biology, or the study of interacting sets of populations ("population of populations"), is gradually deviating from the initial concept of an ideal metapopulation as defined by the founders of the field (review in Hanski and Gilpin, 1997). Scientific concepts, especially those related to Darwinian evolution, must (and do) evolve to adjust to new developments.

The impact of quasispecies for a perception of RNA viruses as highly dynamic, unfixed genetic entities—a perception that, interestingly, was not achieved under the influence of previous models of mutation and selection in populations, as delineated by classical population genetics—provides an additional argument in support of the notion that quasispecies exhibited novel features that suited the conceptual needs of virology. One of the novelties was the emphasis on mutant generation. RNA viruses share with the primordial replicons discussed in Chapters 3-5 the highly error-prone replication (Chapters 4-6). Furthermore, the formulation of quasispecies concepts for RNA viruses coincided in time with an explosion of sequence information as a result of widespread application of rapid methods of nucleotide sequence sampling. Unconceivable just a decade earlier, in the late 70s it then became possible to probe nucleotide sequences of even individual clones from single viral isolates: there was a perfect match of model and of experimental findings emphasizing extreme mutation rates.

Complexity

Complexity is a concept of increasing impact on a considerable number of fields from physics and biology to economy and history. It is a concept with many facets and approaches. It has been viewed as a tendency of systems with numerous components to reach complex interacting structures and behavior, or critical, self-organized states, away from equilibrium (as general introductions see Nicolis and Prigogine, 1977; Bak, 1996; Goldenfeld and Kadanoff, 1999, and for a more detailed description of concepts see Cowan et al, 1994). Biological systems can be regarded as extremely complex systems, their behavior arising as an emergent property from multiple, much simpler, elementary components. Complex systems are highly dynamic and unpredictable: they are governed by unpredictable jumps in behavior. Regarding viruses, simple components underlying their behavior are a rather short polynucleotide chain, mutations, nucleotide-protein interactions, etc. Yet plagues may at times result as a complex output of basic replicative properties of a pathogenic agent.

With a much more modest aim of describing viral quasispecies, complexity can have two distinct meanings. One, used in previous Chapters of this book, is the amount of encoded information included in a viral genome (regulatory regions and open-reading frames). In the absence of redundant information, complexity can, for most comparative purposes, be considered equivalent to the number of genomic nucleotides. A second meaning of complexity takes into account variations in nucleotide sequence when individual genomes of a viral quasispecies are considered. We can assume that in a mutant spectrum each individual genome harbors a common core information shared with the others, which is needed to perform the basic replication functions, those outlined in Chapter 2. Yet, individual genomes acquire mutations which can be calculated in the form of an algorithmic complexity (as an introduction to algorithm complexity see Gell-Mann, 1996 and Simon, 1996; its possible application to describe viral populations was discussed by Domingo, 1999). The algorithmic complexity of a perfectly homogeneous collection of viral genomes, or, paradoxically, a collection of random sequences of

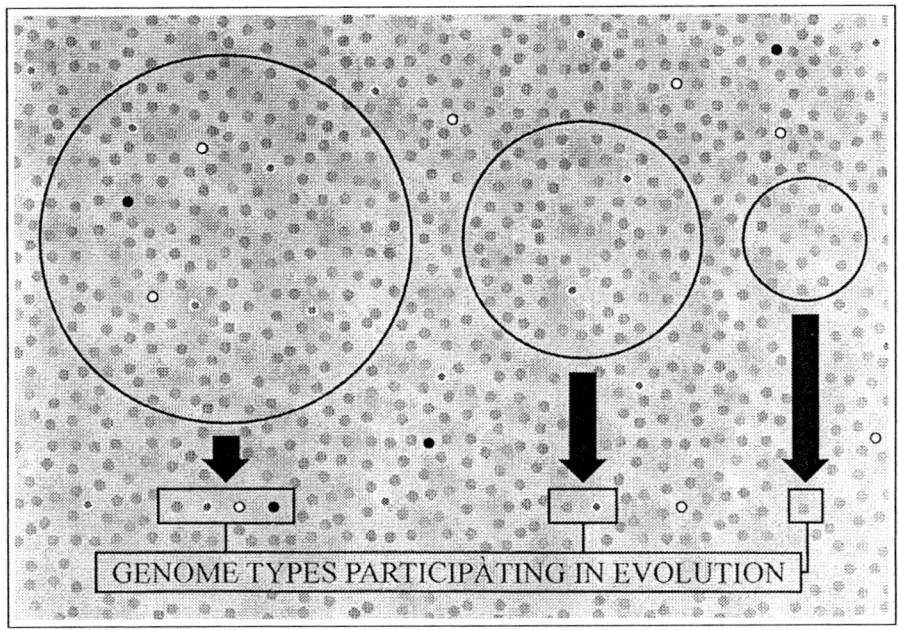

Fig. 8.2. Population size and the evolutionary potential of phenotypic variants of an RNA virus. For a given phenotypic trait variants showing a deviation in behavior are depicted by the low frequency symbols (black, white or dark grey dots). The population size (depicted as circles of different size) that participates in an infection (in cell culture, upon natural transmission of the virus, etc.) determines whether low frequency phenotypic variants will or will not participate in the evolutionary process.

the same length, is far lower than the complexity of the same genomic collection incorporating scattered random mutations. Virologists estimate complexities by comparing mutation frequencies (Chapter 6) or the Shannon entropy (Volkenstein, 1994; Pawlotsky et al, 1998). These calculations, however, do not take into consideration the dynamic component of evolving quasispecies, with the continuing process of mutant generation competition and selection (Chapters 6 and 7; see also Domingo et al, 1998).

In the calculation of mutation frequencies in mutant distributions, population size does not have an influence since the same mutation frequency can be obtained by sequencing 10,000 or 100,000 nucleotides of a mutant spectrum. However, population size may be a critical determinant for the evolution of a quasispecies, thereby introducing a new element of complexity into the previous concept of algorithmic complexity. Examples which refer to viral phenotypes discussed in Chapters 6 and 7 may clarify this point: Any phenotypic variant (antibody or CTL-escape mutant, host-range mutant, etc.) generated at a frequency of 10^{-6} will be found with very high probability when the infecting viral dose is $>10^8$ infectious units, whereas when the infecting population is 10^6 infectious units or fewer, the behavior of the virus regarding possible expression of phenotypic traits will be highly unpredictable (Fig. 8.2).

An important property of complex systems which has been documented for quasispecies is the presence of a molecular memory of past evolutionary events imprinted in the mutant spectrum of viral quasispecies (Ruiz-Jarabo et al, 2000; review in Domingo, 2000). Current observations on RNA virus behavior at the population level suggest that rapidly evolving viruses may become excellent models for studies of molecular evolution as well as for some aspects of the newly developing field of complexity. Research along these lines is currently in progress.

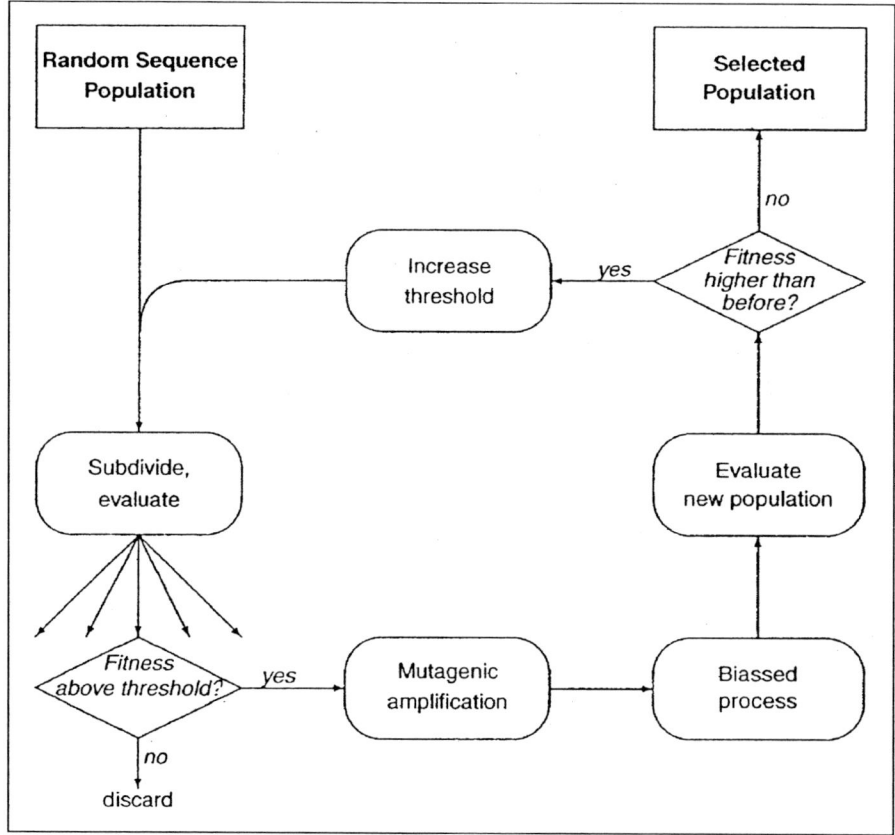

Fig. 8.3. General flow diagram for evolution experiments using artificial selection by a biased process.

RNA Viruses and Evolutionary Biotechnology

A new field of research of increasing impact, and which exploits Darwinian evolutionary principles in vitro is "Evolutionary biotechnology" (Schuster, 1996). Natural biopolymers, in particular nucleic acids and proteins, have an impressive potential to perform or catalyze a rich repertoire of chemical reactions. Only a negligible part of this potential has been tapped for technological innovation. Evolutionary biotechnology aims at the design and synthesis of new polymers endowed with preselected biological and chemical properties. This is achieved to a great extent by the application of replication, competition and selection, as described in preceding Chapters. Evolutionary biotechnology applies the Darwinian principles of generation of diversity and selection to derive molecules with predetermined properties. The latter are improved by serial amplification and selection cycles (Fig 8.3). This principle can best be applied to RNA or DNA since these biopolymers can undergo error-prone replication.

Large libraries of up to 10^{15} molecules can be created from which nucleic acids with new properties can be selected (e.g., "aptamers" that bind ligands with high affinity and specificity, or "ribozymes" that catalyze certain chemical reactions), as is documented in the next sections. Variation can be readily attained with nucleic acids, but not directly with proteins. Therefore, one of the current challenges of evolutionary biotechnology is to design procedures for coupling nucleic acid amplification with its expression into protein and with evaluation of protein function. The availability of variant forms of nucleic acids and proteins may also constitute a tool to approach the folding problem, the ability to predict structures which eventually may give essential hints on how to perform a desired biological function (Rost and Sander, 1996; Leclerc et al, 1997; Schuster et al, 1997). Biological diversity can also be generated in vivo by means of display libraries with biological vectors (phage, viruses, plasmids). Chemical diversity can be achieved by synthesis of peptide or oligonucleotide libraries. Evolutionary biotechnology is a part of the broader field of "Molecular Biotechnology" in which not only evolutionary concepts but also other areas of molecular biology, biochemistry and biophysics are exploited for technological purposes (Schuster et al, 1998). Many DNA and RNA viruses and other RNA genetic elements such as satellite RNAs have been modified to be used as vectors for the amplification of foreign genes, gene targetting, cell-specific delivery, and gene expression.

Synthesis of Biopolymers

It has been recognized recently that molecules with new properties can be constructed with the help of "combinatorial chemistry" where building blocks with suitable characteristics and sidegroups are condensed step-by-step in a specific sequence to yield a polymer. However, the exhaustive exploration of the sequence space for suitable candidates becomes more and more difficult as the numbers of different monomers and their chain lengths increase. Sophisticated screening methods are required to select for macromolecules with the desired properties.

In the center of interest of combinatorial chemistry are oligopeptides and oligonucleotides because the methods employed in their chemical synthesis can be supplemented with enzymatic reactions and also because they can be used for biological interactions. Peptides can mimic some antigenic sites of pathogens and may find application in the formulation of vaccines. Oligonucleotides serve to identify genes and even genomes by means of highly specific hybridization reactions. The solid phase synthesis methods of oligopeptides and oligonucleotides have been improved and highly automated so that large amounts of oligomers up to chain lengths of 100 residues are readily available. Longer chains can be obtained, but their purity often does not suffice for many applications. Side products with shortened sequences accumulate because the reactions of coupling and deblocking do not proceed to full completion. A capping step that blocks unreacted groups from being elongated in the next cycle has reduced the side products, but the maximal fidelity obtained by chemical methods is still no match for the fidelities obtained by enzymatic methods.

For synthetic oligonucleotides or oligopeptides, parallel synthesis methods have been developed that produce an array of different sequences arranged on a chip (Fodor et al, 1991) or membrane. Large libraries of oligomer sequences can be screened simultaneously for desirable properties by such a high-throughout method. Hybridization with species-specific oligonucleotides is a fast and reliable method for unequivocally identifying an organism or a virus within a biological sample which may include a mixture of various organisms. The number of hybridized genomes required to give a positive signal depends on the method, but is still above 10^5 copies. A promising development is the DNA-silicon chip method where the genetic diversity of an organism can be accessed by hybridization to high-density arrays of oligonuleotide probes (Lipshutz et al, 1995).

For exploring a fitness landscape in sequence space, it is impossible to synthesize all members of the sequence space of oligonucleotides or peptides, even at moderate chain lengths and with large-scale parallel-synthesis devices. Sophisticated strategies must be used to find successful sequences in an enormous number of blanks. Two general strategies can be chosen: "rational" and "irrational" design. Rational design is feasible for well-understood systems. It is used to optimize or alter the properties of oligomers that are known to fulfill some required activity.

Position-directed mutations are introduced to modify the sequences, the properties of the mutants are evaluated, and the most successful ones are selected.

If one wants to design a novel function there is no well-characterized starting point, and an "irrational design" approach has to be taken. This simple method, used for a long time in screening chemicals for their possible use as drugs, has regained interest for the systematic creation of new compounds by combinatorial chemistry: A large number of different sequence combinations is synthesized and screened for the desired property. Once a succesful sequence is found by chance, its neighborhood in the sequence space is explored systematically to locate the highest peak in the fitness mountain. In principle, irrational design does not depend on prior knowledge of any correlation between sequence and function. In reality, however, the probability of hitting a successful sequence by a random shotgun experiment is so low that a mixed approach is preferable: whatever information is available to restrict the shotgunning to a narrow field is used to increase the probability of success. The exploration of sequence space can be sped up by genetic computer algorithms that reduce the number of trials with minimal risk of getting trapped in local optima (Forrest, 1993).

Peptide libraries have been screened for binding to antibodies. A fairly representative exploration of the fitness space (binding strength to the antibody being the measure of fitness) has been possible. As expected, the natural antigen and the closest mutants thereof form a mountain in the landscape. However, other sequences not related in their sequences to the natural antigen were also found (Li et al, 1998) but bind strongly to the antibodies. Together with their mutant spectrum they form another, independent mountain in the fitness landscape.

Recombinant DNA Techniques

Larger polymers could be synthesized in principle by an analogous combinatorial synthesis from oligomer building blocks with defined sequences. Besides the limited purity of synthetic oligomers, the enormous number of components required to allow all possible sequence combinations reduces the suitability of such an approach for irrational design. Synthesis of polynucleotides and polypeptides in the chain length range of genes or gene products is still not a practicable task. However, there is an impressive number of biochemical tools for in vitro production of recombinant DNA by which DNA pieces derived from different organisms or viruses can be specifically combined or specifically altered (Sambrook et al, 1989). The enzymes used involve the DNA/RNA-modifying enzymes described in Chapter 3 as well as sequence-specific DNA restriction endonucleases and DNA ligases that permit the dissection of a gene (or even a genome) into specific pieces which can be ligated with a specific plasmid or viral vector. The new combination can be amplified and expressed in vivo by transformation of a host organism with an expression vector containing such a DNA insert. The new genes can be sequenced and the properties of the gene products determined.

RNA genome manipulation is much more difficult. In most cases production of recombinant RNA in vitro involves retrotranscribing the RNA into DNA, performing the recombinant DNA techniques with this cDNA, and then retranscribing the recombinant DNA into RNA with the help of a suitable in vitro transcription system.

Transformation with a vector changes the genome of the host and thus its phenotype. The gene introduced may come from another organism, forming a genetic "chimera". Chimeric DNAs are finding many applications in the characterization of genes, the synthesis of new biological reagents, and in the modification of cells, with a potential use in gene therapy. At present, a spontaneous gene defect often becomes evident only after most of the developmental program has already been executed, and successful gene therapy requires the transformation of a subset of somatic cells of the organism. Gene therapy introducing modified genes may improve the survival of an organism under stress conditions (Encell et al, 1998, 1999).

RNA Virus Vectors

DNA viruses (bacteriophages lambda, M13; adenoviruses, baculoviruses, herpesviruses and poxviruses, in particular vaccinia virus, parvoviruses and papillomaviruses, among others) and retroviruses have been extensively used as vectors for amplification and expression of foreign genes (Coen and Ramig, 1996). Despite the genetic instability usually manifested by RNA replicons when undergoing serial rounds of replication, a number of RNA viruses and defective RNAs have been also engineered as vectors. First, infectious clones were constructed for bacteriophage Qβ, poliovirus and many positive strand RNA viruses which represented the introduction of reverse genetics in the field of RNA virology (Taniguchi et al, 1978; Racaniello and Baltimore, 1981; review in Gromeier et al, 1999). Then reverse genetics was extended to negative strand RNA viruses (García Sastre and Palese, 1993; Pattnaik et al, 1992; Schnell et al, 1994; Radecke et al, 1995). Animal and plant RNA viruses and defective RNAs that require a helper virus for replication were engineered to express foreign proteins (Luytjes et al, 1989; Levis et al, 1987; Collins et al, 1991; Hahn et al, 1992; Burgyan et al, 1994; Spielhofer et al, 1998). These procedures allow expression of foreign proteins inside infected or transfected cells, or formation of viral particles that include a foreign protein in their structure. In the latter case, a viral gene can either be replaced by a functional analogue from another virus (e.g., a nucleocapsid protein or a surface glycoprotein) or a surface protein may be engineered to include foreign epitopes in a chimeric form, often for the purpose of designing an antiviral vaccine. An example was the construction of a poliovirus chimeric construct exposing antigenic sites from the human immunodeficiency virus (Evans et al, 1989). In a recent study, the negative strand unsegmented measles virus was engineered to express the distantly related surface protein from vesicular stomatitis virus (Spielhofer et al, 1998). These chimeric viruses protected mice against lethal doses of wild type vesicular stomatitis virus.

Problems that might arise in the use of RNA virus vectors are the introduction of point mutations in the foreign gene resulting in the expression of altered proteins, or the loss of the foreign insert through recombination. These events might occur if fitness of the chimeric constructs is lower than the fitness of unmodified vector virus, and if the number of replication rounds is not maintained to a minimum (Chapter 7). Regarding the evoking of immune responses by antigenic proteins or epitopic domains, it is unlikely that RNA virus vectors may represent any substantial improvement over classical vaccines. The problem of ensuring a broad immune response, similar to the response evoked by the authentic viral pathogen [the hallmark of an efficient vaccine (Kaslow and Evans, 1997)] may require the engineering of RNA vectors capable of expressing in a functional way multiple B- and T-cell epitopes (see following sections of this Chapter).

Antisense RNA

Nearly as important as adding a gene to a complex DNA is to be able to suppress expression of one or several genes (Roy and Harris, 1994). Particularly important is an antiviral strategy consisiting in knocking out vital functions in the infection cycle of a virus (Koschel et al, 1995). A plausible approach is the inactivation of a mRNA by a so-called "antisense RNA", i.e., a segment of RNA complementary to an important region of a mRNA. Double strand formation between the mRNA and the antisense RNA can abolish ribosome binding or derail translation of the mRNA.

The practical application of this plausible approach is not easy. As discussed in the previous Chapters, RNA viruses produce their own antisense RNA, in most cases without noxious effect on the reproduction of the viral RNA. At least for prokaryotic viruses it is quite clear that a double-stranded RNA represents a dead end; it can be neither translated nor replicated, and there is no biochemical pathway to melt a double strand into single strands. Apparently, double strand formation does not take place to an extent that may inhibit RNA amplification. In vitro studies of double strand formation have shown that it is a slow process. High sequence complexities, strong secondary structures or the presence of protein factors may slow down double strand formation.

The efficiency of an antisense RNA thus depends on many factors: (i) it must be present in high concentrations, (ii) it should be rather short, (iii) should not contain secondary structures that might impair the interation with target RNA, (iv) should be complementary to a readily accessible region of the mRNA, (v) should knock out an essential binding step of the mRNA and (vi) should be imported efficiently into the target cells, and (vii) its half-life should be sufficiently large as to exert its inhibitory action, ensuring resistance to ribonucleases. For this reason, nucleotide analogs with altered backbones are often used in the chemical synthesis of antisense RNA (i.e., by protecting the 2'-OH of the ribose moieties by chemical modification, or by replacing them by H or F). Also, the phosphodiester backbone may be partially or entirely replaced by a peptide backbone. Several techniques have been developed to introduce the antisense RNA into the cell, e.g., by including the RNA into vesicles or microspheres formed by lipid bilayers. Antisense RNA technology has to be carefully adapted to each application in order to be effective; nevertheless, substantial progress has been achieved for some applications (Wagner and Simons, 1994).

Molecular Amplification Systems

Synthetic methods of finding biopolymers with novel properties are generally limited to rather short chain lengths, too short for efficient catalysis. Successful in vitro design of new catalysts requires Darwinian evolution, involving mutation, amplification and selection (Eigen and Gardiner, 1984; Kauffman, 1992).

The first molecular amplification system was the RNA replication by the RNA replicases of leviviruses (Chapter 3). In principle, a single chain of a replicable RNA can trigger an amplification avalanche to produce in a rather short time a macroscopically detectable amount of RNA which may be used as an extremely sensitive reporter molecule (Chu et al, 1986). A serious problem for the amplification of any RNA is the high specificity of the viral replicase. This problem has been overcome partially by inserting a sequence into an appropriate replicon (Lizardi and Kramer, 1991; Wu et al, 1992). Inevitably, the modification of the replicon diminishes its fitness, and there is a strong selection pressure to reoptimize the recombinant RNA by changing or deleting the insert. Furthermore, unmodified replicons must be absent, because they would be rapidly reselected. The smallest contamination of equipment, reagents and room with unmodified, standard templates strongly interferes with the amplification of the target replicon (Munishkin et al, 1991; Biebricher et al, 1993). It is thus imperative to monitor that the replicon being amplified maintains the desired insert. The presence of a target sequence inserted into the RNA replicon can be verified by hybridization (Tyagi and Kramer, 1996) or nucleotide sequencing. Because of likely unwanted evolutionary events, suitable amplification factors are limited to a few orders of magnitude. The high specificity of the replicases used severely restricts the sequences that are tolerated within a replicable species. While replicase catalyzed reactions are indeed used for enhancing the sensitivity of assays, e.g., of probes hybridizing to a target sequence, its use has not found the general application spectrum of other amplification methods.

The search for more generally applicable molecular amplification systems has led in the past decade to novel, highly successful amplification methods. The most important is the polymerase chain reaction (PCR; Mullis et al, 1986; Saiki et al, 1988). A successful amplification requires recycling of the template (see comparisons in Table 3.2, Chapter 3). If a polymerase is not able to recycle the template, the products of the synthesis must become new template molecules. PCR amplifies double-stranded DNA by melting it thermally into single strands and then extending oligonucleotide primers corresponding to the 5' termini of both strands to full strands by a DNA polymerase (Fig. 8.4). Instrumental for the high success of the PCR reaction was the use of thermostable DNA polymerases that withstand the high temperature needed for melting the DNA double strands (Saiki et al, 1988). It allowed the automation by which a very accessible apparatus provides for a programmed temperature cycle: a high temperature at which the DNA melts, a lower temperature for oligonucleotide primer hybridization to DNA, and the optimal temperature for the DNA to complete the elongation reaction. Iterating these reaction steps doubles the DNA concentration in each cycle, i.e., the DNA concentration increases exponentially.

Automation of the thermal cycle and the use of a thermostable DNA polymerase have made PCR a highly useful tool in biotechnology (Schober et al, 1995). The method is rather general: provided that some conditions are met in selecting primers, reaction times and temperatures, almost any DNA sequence can be amplified. Improvements of the enzymes used for amplification have increased the sensitivity, the fidelity and the size of the DNA that can be amplified. The coupling of retrotranscription and PCR (RT-PCR) has allowed analysis of RNA viral populations, in particular viruses which cannot be subjected to biological cloning, as discussed in previous Chapters. PCR methods, coupled to automated sequencing (cycle-sequencing), have substantially reduced the time required for the determination nucleic acid sequences.

An extension of this system amplifies RNA by a network of transcription and retrotranscription: The RNA template is transcribed into cDNA by reverse transcriptase. The DNA strand is released either by heating or by digesting the RNA template with RNase H. The resulting single-stranded DNA is completed to a double-stranded cDNA by a DNA polymerase using a primer containing a promoter sequence followed by the 5' terminal sequence of the RNA template. The resulting double-stranded DNA produces some 100 strands of RNA by RNA polymerase. The reaction is called self-sustained sequence replication (SSR) (Kwoh et al, 1989; Guatelli et al, 1990). While this experimental procedure is more complicated than direct RNA replication, there is much less constraint on the RNA sequence to be amplified. The temperature cycle method of RNA amplification has been mostly replaced by its isothermal alternative using RNase H (Fig. 8.5). The method has the advantage that costly automates are not required and that single-stranded RNA of only one polarity is produced; interference by the antisense strand is avoided. Disadvantages are the limitation of the practicable amplification factors by the rapid emergence of side products (Ellinger et al, 1998) or of molecular parasites (Breaker and Joyce, 1994), and the limitation of the RNA chain lengths that can be amplified to a few hundred bases. Therefore, viral RNA cannot be amplified by this method.

Selection of RNA with a Function

In the last decade, many biopolymers with new or altered binding or catalytic properties have been developed. Essential for this approach was the enrichment of the advantageous sequences within a population, e.g., by selective amplification by a reliable amplification method. The invention of amplification methods for RNA has thus given RNA a leading edge over proteins despite the higher efficiency of proteins for specific binding or catalytic activity. However, while RNA replication by viral replicases is a comparatively simple reaction, it is too discriminatory for particular sequences to be generally applicable (Biebricher and Gardiner, 1997). At the other extreme, PCR is rather sequence-independent, but double-stranded DNA is unable to build the compact tertiary structures that are necessary for a specific function. The

Fig. 8.4. The polymerase chain reaction (PCR). A solution containing the DNA template, primers containing the 5'-termini of the two complementary DNA strands, the four deoxyribonucleoside triphosphates and a heat-stable DNA polymerases is undergoing a programmed temperature cycle: At the high temperature, the DNA is melted into single strands, at the low temperature, hybridization of the primers with the single-stranded template takes place, and at the optimal temperature for the DNA polymerase the primers are completed to full length double strands. At each cycle, the DNA concentration doubles.

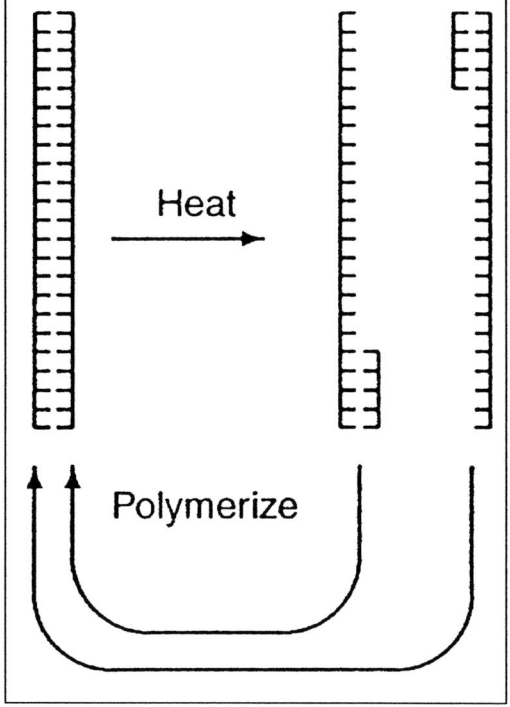

discovery of single-stranded DNA functionally competent to cleave RNA is only recent (Santoro and Joyce, 1998). Interestingly, in the few cases where single-stranded DNA and RNA species have been selected with equivalent functions, no sequence homology was found, indicating that structural elements other than base-pairs must contribute substantially to the function (Joyce, 1998).

Starting from a random RNA library (Biebricher and Orgel, 1973), a number of highly specific RNA "aptamers" have been selected that bind a low molecular weight substrate molecule with high specificity. The usual technique is systematic enrichment of ligands by exponential amplification (SELEX) (Tuerk and Gold, 1990). Test-tube experiments involving cycles of amplification, mutagenesis and selective binding to immobilized substrates have led to optimization of RNA sequences that bind to specific targets (Ellington and Szostak, 1990; Niewlandt et al, 1995). The specificity is particularly high for substances which can form hydrogen bonds and base stacking to nucleotides of the RNA; RNA aptamers have been found to bind highly specifically to viral proteins (Convery et al, 1998).

The highly specific binding of substrate molecules is one of the prerequisites for catalysis. In contrast to proteins, however, RNA lacks chemical side groups. The detection of "ribozymes", RNA with catalytic properties by Cech (Kruger et al, 1982; Zaug et al, 1983) and Altman (Stark et al, 1978; Guerrier-Takada et al, 1983) was thus a surprise. The natural ribozymes they found participate in phosphoester transfer and phosphoester hydrolysis reactions. Furthermore, it was found that the RNA moiety of organelles participating in gene expression, e.g., of ribosomes or spliceosomes, contribute considerably to the enzymic reactions catalysed by them (Crick, 1968; Dahlberg, 1989; Noller, 1993).

The facile amplification of RNA has stimulated an active search for RNA that catalyses novel biological or nonbiological reactions. Selection from randomized or partially random-

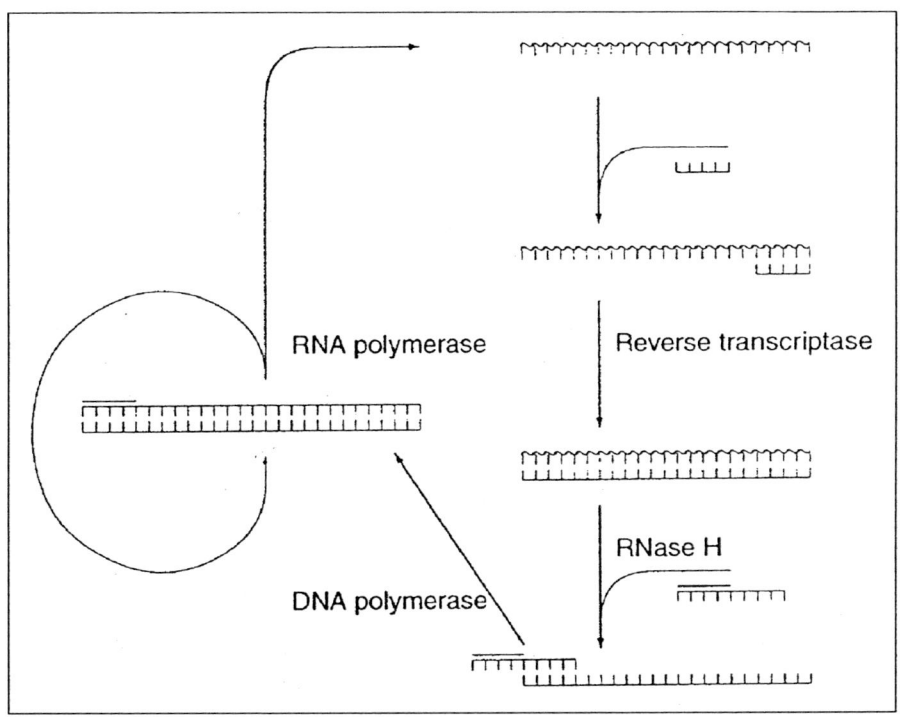

Fig. 8.5. RNA amplification mechanisms by the 3SR method. Primed retrotranscription of the RNA (wavy line) results in a DNA: RNA hybrid strand. The cDNA strand (straight line) is liberated by heat or by digestion of the RNA moiety with RNase H. Using a primer containing a T7 promoter sequence, the single DNA strand is completed to the DNA double strand. T7 RNA polymerase uses this template to synthesize 300-1000 copies of the input RNA. The RNase H version proceeds isothermally and does not need a programmed temperature cycle. Most reverse transcriptases include the activities of RNase H and DNA polymerase, so that addition of reverse transcriptase, T7 RNA polymerase, the four ribonucleoside triphosphates, the four deoxyribonucleoside triphosphate, RNA template and the two primer suffices.

ized sequences, followed by iterative cycles of amplification and selection (Szostak, 1993), has led to the creation of ribozymes with an amazing repertoire of reactions. Among others, ribozymes serving as RNA ligase (Cuenoud and Szostak. 1995; Chapman and Szostak, 1995), aminoacyl-RNA synthetase (Illangasekare et al, 1995), peptidyl transferase (Zhang and Cech, 1998), polynucleotide kinase (Lorsch and Szostak, 1995), RNA polymerase (Doudna and Szostak, 1989; Doudna et al, 1993) and alkylase (Wilson and Szostak, 1995) have been selected. Most novel ribozymes were created entirely de novo, but it was also possible to optimize existing ribozymes or to extend their catalytic repertoire by coupling the desired activity with a selective advantage in the amplification procedure (Joyce, 1992; Beaudry and Joyce, 1992). Even DNA single strands have been found to catalyze specific reactions (Joyce, 1998). Since RNA lacks groups capable of performing effective acid-base or redox reactions, modification of nucleotides—as occurs naturally in tRNA modification—or incorporation of nonnatural nucleotide analogs (Piccirilly et al, 1990), in particular nucleotide coenzymes, offer further possibilities to extend the repertoire of ribozymic functions. Ribozymic self-incorporation of a coenzyme has already been realized (Breaker and Joyce, 1995).

Protein Design

Despite the progress of ribozyme technology, there is no doubt that proteins are generally more suitable than nucleic acids for catalysis. The large factors of speeding up reactions effected by protein catalysts—factors of 10^{10} are not rare—can not be obtained with ribozymes, and ribozymic reactions often require highly specialized conditions.

Design of large proteins has proven a difficult task, not only because the sequence space grows rapidly to hyperastronomical proportions (Chapter 5), but also because the rational design of a protein structure from a completely new sequence is still largely unfeasible. However, it has been possible to explore the neighborhood of a successful protein in the sequence space and to investigate a range of structures and functions (Fersht et al, 1984). Evolutionary techniques, e.g., selection after partial randomization, can be employed to alter the properties of enzymes (Bornscheuer 1998, Loeb, 1996; Munir et al, 1993), e.g., to make them more thermostable (Giver et al, 1998). If the structure of the master sequence is known, the predictive power of calculations determining the structural consequences of a mutation is satisfactory. As predicted by the theory, a large number of mutants with no or nearly no difference in biological activity, e.g., an enzymic activity, are found among the nearest neighbours in sequence space. Only a few single amino acids exchanges abolish the enzymic activity, e.g., when the amino acid affects the active center of the enzyme or when the exchange disrupts an important structural element. However, there is an apparent discrepancy: Together with the silent mutations, a much higher number of mutants of an organism or a virus would be expected to be neutral than is actually observed. The fitness of a mutant in vivo depends also on factors other than enzymic activity, e.g., interactions with other components of the cell: The global fitness of the whole organism is evaluated, and deviations in the enzymic activity of a protein usually have less effect than the interplay with the large number of other molecules in the community of the cell.

A truly innovative in vivo evolution of an entirely new function has not yet been observed. As discussed in Chapter 4, organisms rapidly adapt to new environments, but they do so in regulating up and down already existing activities rather than inventing new gene products. On the other hand, Clarke (1986) succeeded in selecting bacteria that were able to metabolize certain sugars that could not be utilized by the precursor strain. An analysis of the mutations that led to this new activity revealed, however, that, rather than creating a new activity, the high substrate specificity of metabolic enzymes was relaxed to accept additional metabolites.

Evolutionary success is the result of a delicate balance of different interactions in a very complex metabolic network which is likely to be disturbed by the introduction of a foreign gene. Only when the selection pressure gets extreme, e.g., by providing a deadly poison in the medium, organisms can be forced to either accept the gene to neutralize the poison or to perish. It has been possible by gene transfer followed by evolutionary adaptation to select enzymes with novel properties (Christians and Loeb, 1996) or genetic expression elements like promoters (Horwitz and Loeb, 1988), but profound alterations to generate highly fit, new cell types or organisms remains an extremely difficult undertaking.

Catalytic Antibodies

For the selection of proteins binding to almost any chemical compound, nature has invented a highly successful evolutionary system, the humoral immune response of higher animals designed to fight off invading parasites. From the beginning scientists were puzzled by the enormous diversity of the immunoglobulins: antibodies could be raised against any chemical "hapten" group, even against unphysiological compounds (Edelman, 1970). A mammalian genome is too small to possibly provide genes for all the immense repertoire of antibodies. The diversity to generate antibodies, brought about by the combinatorial rearrangement of multiple genetic elements aided by RNA editing, is so high that an organism cannot possibly keep all antibodies in the required concentration in stock. On demand, the clonal selection caused by binding of an antigen to a B-type immune cell triggers growth and division of the cell (Burnet, 1959; Jerne, 1971), resulting in synthesis of large amounts of the required antibodies. The technique to generate monoclonal antibodies (Kohler and Milstein, 1980; Milstein, 1980) revolutionized the technical exploitation of antibodies. An ingenious method employs the production of antibodies against any structure for generating novel catalytic proteins: It is known that enzyme binding of substrates distorts its geometry towards the transition state (Chapter 3). If a susbstance that has an analogous structure to the transition state is synthesized, antibodies recognising this structure have the potential of catalysing the reaction from this transition state. This technique has been highly successful: catalytic antibodies for a number of reactions have been produced (Lerner and Tramontaur, 1988; Schultz, 1989; Schultz and Lerner, 1993), including reactions that are not used in cell biochemistry. Even though their catalytic efficiencies do not reach the efficiencies of enzymes that have been optimized by a long evolutionary process, their stereospecificity makes them excellent synthetic tools. Attempts have been made to devise combinatorial strategies to explore repertoires of catalytic antibodies (Iverson et al, 1989).

Phage Display

In vitro translation is not very efficient, and the coupling of in vitro replication and in vitro translation to select proteins with required properties is still a very difficult task. More successful has been the use of recombinant DNA techniques in vitro, coupled to in vivo translation. One powerful technique is the display of a peptide on the surface of a virus by fusing a structural viral gene with the required sequence. The technique has been developed using phage M13 (Parmley and Smith, 1988; Scott and Smith, 1990; Smith and Scott, 1993). While in principle the major coat protein of M13 could be used as the carrier of the displayed peptides, the presence of a few thousand copies of the foreign sequence causes severe restrictions on the possible peptide sequences. Therefore, the peptide library has been fused to gp3, the protein responsible for binding of the phage to the F-pilus of the host cell. Even though the fusion with the displayed peptide is at a domain involved in host recognition, sequence restrictions are not severe. Peptide libraries can be displayed on the surface of a large phage population. Artificial selection can then be used to isolate phages containing peptide inserts with the best performance in a diagnostic reaction, and the selected clones can be further amplified by successive infection cycles, since the information for the peptide production is contained in the pertinent phage genome. Phage display of single-chain antibodies (Griffiths, 1993) has opened a way to make large libraries of antibodies and select the ones with the desired properties without immunizing an animal, avoiding thus a procedure with a rather unpredictable outcome. Phage display and the enrichment of successful phages by binding them to an antigen-bound surface (a technique called biopanning) have been successfully used in the selection of catalytic antibodies.

However, some steps in the phage display technique need to be improved. While introducing randomized sequences into the genome is readily possible, the limiting step is the transformation of bacteria to produce the phages. At least 5 orders of magnitude in the phage population are lost by this quite inefficient step. Furthermore, adaptation by phage multiplication and selection cycle are too slow because M13 DNA replication is too accurate. Recombination has to be suppressed by using appropriate host strains to avoid deletions of the inserted material. Mutations have to be introduced in vitro, e.g., by DNA shuffling (Crameri et al, 1998; Harayama, 1998) and after each step another inefficient transformation step is required.

The high adaptation potential of the RNA phages can be used for phage display. The manipulation of the genome itself can be easily performed at the cDNA level. The RNA phages are liberated from cells transformed with the cDNA of the viral genome (Taniguchi et al,

1978), but they could also be produced in principle by in vitro transcription followed by phage assembly. Phenotypic expression of the RNA requires a full infection cycle.

Altering Mutation Rates

The rate of evolution is strongly dependent on the mutation rate, and it would be of great value to have the ability to artificially adapt mutation rates to experimental needs. Fifty years ago bacterial "mutator" strains were isolated that apparently displayed an enhanced error rate. Most of these strains are defective in genes involved in DNA repair. In previous Chapters we have seen that the fidelities of RNA viruses with similar amplification strategies may vary within nearly two orders of magnitude and that mutations affecting error rates have been introduced in some replication enzymes. However, dramatic effects on copying fidelity have not been reported.

As also discussed in previous Chapters, replication fidelity is dependent on the conditions (presence of nucleotide analogs, metal ions, bias in the monomer composition, etc.) Procedures have been developed for mutagenic PCR amplification of DNA (Meyerhans and Vartanian, 1999). As discussed in the next sections, increases in mutation rates induced by nucleotide analogues constitutes a promising, new antiviral strategy which exploits quasispecies dynamics to the detriment of virus survival.

Quasispecies and Viral Disease Control Strategies

For several decades virus evolution was considered to be a field of rather speculative research, quite unrelated to the core issues of virology: the understanding of virus structure, replication, pathogenesis and development of vaccines and antiviral agents to eradicate viruses and the diseases they produce. The picture has changed dramatically, and quasispecies and rapid RNA virus evolution are now intimately linked to failures in viral disease control and prevention (Domingo, 1989; Domingo and Holland, 1992; Duarte et al, 1994; Novella et al, 1995; Levin et al, 1999). Critical to the difficulties for the control of viral disease has been the design of antiviral strategies using single and fixed elements (one antiviral agent, a vaccine based on a single synthetic peptide) to control highly dynamic viral quasispecies. Major problems and possible solutions are summarized in Table 8.2. Implementation of combination antiviral therapies and the recognition that vaccines must be multivalent are important steps in adapting disease-control strategies to the population complexity of the pathogens. It is now remarkable to think that just one decade ago clinical trials involving administration of mixtures of antiviral drugs were considered rather unscientific because they violated the basic principle that the effect of one variable should be tested at a time.

But are the new strategies summarized in Table 8.2 sufficient to respond to the amazing genetic plasticity of viruses? As we will discuss in a next section, not only the intrinsic adaptability of RNA viruses, but also several forms of environmental modification, tend to perturb viral populations, resulting in an enhanced probability of emergence of new viral pathogens. This permanent challenge invites exploration of new antiviral strategies, in particular those which take into consideration the highly mutable nature of viral replicons.

New Antiviral Strategies Based on Violation of the Error Threshold

Error-prone replication entails the existence of an error threshold which defines the maximum genetic information that can be maintained stably with a replication machinery endowed with a given copying fidelity (Chapter 4). This concept, first proposed on theoretical grounds (Eigen, 1971; Swetina and Schuster, 1982; Eigen and Biebricher, 1988), has found considerable experimental support. A number of chemical mutagens (5-fluorouracil, 5-azacytidine, ethyl methanesulfonate, nitrous acid) which act either on replicating viral RNA in infected

Table 8.2. Major problems and guidelines for antiviral strategies

PROBLEMS
- Preexistence or selection of antibody- and CTL-escape mutants and drug-resistant mutants in viral populations.
- Attenuation and virulence are not fixed traits.
- Escape mutants may either show high fitness immediately or may readily gain fitness[a].
- Even adequate administration of combinations of antiviral inhibitors may not reach all sites of infection (or reach them with ineffective concentrations).

GUIDELINES
- Vaccines must be multivalent (including multiple B-cell and T-cell epitopes).
- Preference for vaccine development: attenuated > whole-virus inactivated > multiple protein subunits > single protein > synthetic peptides[b].
- Vaccines may have to be periodically updated[c].
- Use of a single monoclonal antibody to suppress viremia should be avoided.
- Antiviral drug therapy should involve combination therapy with several drugs which do not share a common mode of action, nor cross resistance[d].
- Dominance of drug-resistant mutants in pathogen populations should be avoided. This may require temporary shelving of drugs.
- Treatment of individual patients with antiviral drugs should be discontinued when the virus population has acquired drug resistance.
- Suboptimal doses of vaccines or drugs should be avoided.

[a] Fitness gain in connection with antiviral strategies has been reviewed in Domingo et al (1997).
[b] Safety considerations may play a role in vaccine design (proviral integration in live retroviral vaccines, immunopathological sequelae of some forms of inactivated vaccines, etc.). Mixtures of proteins or synthetic peptides representing variant forms of the same antigenic region may increase their effectiveness.
[c] This is presently done to control important diseases such as human influenza or foot-and-mouth disease of animals.
[d] The number of required drugs may depend on the viral population size and turnover of virions and infected cells.

cells or the RNA of viral particles, could increase mutation frequencies at specific sites of infectious vesicular stomatitis virus and poliovirus genomes by a maximum of 3-fold (Holland et al, 1992). Attempts to increase mutation frequencies with higher doses of the mutagens failed, and resulted in significant decreases in viral infectivity. In a similar study with a retroviral vector, the increase in mutation rate induced by 5-azacytidine reached 13-fold (Pathak and Temin, 1992). This slight difference could reflect a lower mutation rate for retrovirus than for riboviruses (Drake, 1993) and a smaller genomic target size for the retroviral construct than for the poliovirus and vesicular stomatitis virus genomes. Increased mutagenesis was later shown to exert negative effects on expected fitness gains of VSV clones (Lee et al, 1997).

In a more recent study, Loeb et al (1999) showed that the mutagenic base analogue 5-hydroxycytidine resulted in loss of HIV-1 replicative potential after multiple passages in human cells. Sequence analysis documented an increase in G → A transitions as expected from the mutagenic action of the analogue. These results with a number of viruses suggest that if drugs could be designed with the ability to specifically increase the viral mutation rates, this would constitute an effective new antiviral strategy to push RNA virus error-prone replication into error catastrophe (Holland et al, 1990; Pathak and Temin, 1992; Domingo et al, 1992;

Domingo and Holland, 1997; Ji et al, 1994; Loeb et al, 1999). If such an approach were to prove feasible, it should find its most significant application for the elimination of riboviruses from infected organisms. This group of nonretroviral RNA viruses has contributed most emergent and reemergent viral diseases (such as Ebola, hantaviruses, and arenaviruses; see next section). In the case of retroviruses, their hiding within the host chromosome as proviral DNA, would probably render them relatively resistant to increased mutagenesis. Indeed, the reversion of some *lacZ* mutants of *Escherichia coli* could be increased several thousand-fold by chemical mutagenesis (Cupples and Miller, 1989) because of the much less error-prone nature of DNA replication/repair. For retroviruses, combined stimulation for provirus expression and lethal mutagenesis could be considered. Therefore, replication of RNA viruses close to the error threshold (Holland et al, 1990) opens the possibility of exploiting their high mutation rates to drive viruses towards extinction.

The Emergence and Reemergence of Viral Diseases

Perhaps one of the areas in which complexity is most obviously manifested is in the emergence of new infectious diseases, a major concern for public health authorities at the dawn of a new century (Lederberg et al, 1992; Scheld et al, 1998). Emergence and reemergence of viral diseases can be influenced by a number of ecological, environmental and demographic factors (Fig. 8.6) (Morse, 1993; Shope and Evans, 1993; Murphy, 1994; Murphy and Nathanson, 1994; Kilbourne, 1994; Mahy, 1997).

In dramatic contrast with optimistic predictions of global eradication of many infectious diseases just a few decades ago (discussed in Chapter 6), there is now an increasing awareness that the fight against infectious diseases is far from over, and indeed may continue indefinitely. The AIDS epidemic, with 40 million infected people worldwide in the year 2000, has unveiled profound weaknesses in our infectious disease control capabilities. Yet AIDS is not the only concern. In the last 15 years more than 40 emergent human viruses have been described (Mahy, 1997). In addition, a number of reemergent viruses are expanding into new geographical locations (for example, Dengue in South America, and raccoon rabies in northeastern states of the US) (Murphy, 1994).

The terms emergence and reemergence require clarification. Obviously, by emergence we do not mean the completely de novo generation of a viral entity, for example, those resulting from intracellular RNA recombination events involving segments of viral and cellular RNAs, which lead to a new combination of functional modules and autonomous replication. In this sense, new viruses must be exceedingly rare. Viral emergence usually refers to a newly recognized viral pathogen which previously did not manifest any of the disease symptoms with which it is now identified. Previously, the same (or a closely related) virus may have been infectious only for some other host species. The term reemergence often refers to a well recognized agent which expands to a new geographical area or displays a significant increase in incidence or disease severity.

Emergence and reemergence of viral diseases are the result of multiple complex factors that extend from the evolutionary potential of the viral pathogens to environmental and technological influences (Fig. 8.6). Of the 42 new, emerging and reemerging human viral pathogens listed by Murphy (1994), only 5 are DNA viruses, and one of these is hepatitis B virus which uses RNA as a replicative intermediate (Chapter 2). This clear bias in favor of "new" RNA viruses is maintained when emergent and reemergent viruses of animals and plants are considered. The rapid evolutionary potential of RNA viruses, their continuous exploration of sequence space via mutation, recombination and genome segment reassortment, have been extensively emphasized in this book (see the quantitative listings in Table 6.3 of Chapter 6 as support for such evolutionary potential). Effects of genetic change in viruses, which favor emer-

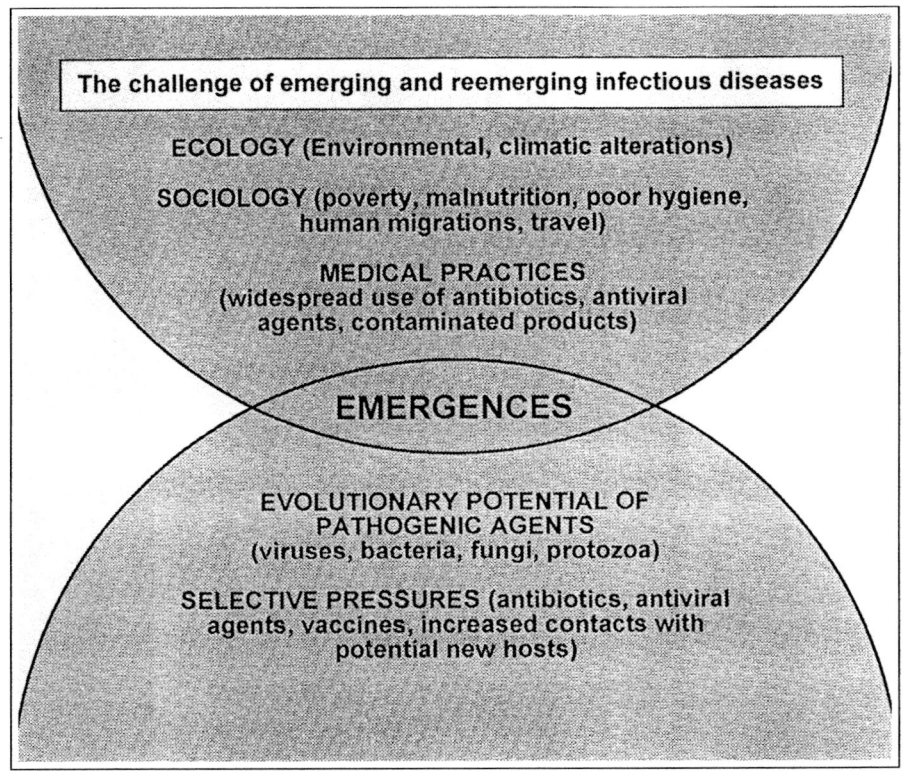

Fig. 8.6. A complex stray of disparate influences may participate in the emergence of new viral disease. Additional information and relevant references are given in the text.

gence, reemergence or a change in virulence, have been documented for influenza viruses (including fowl plague), western equine encephalitis, rubella, canine parvovirus disease, and subacute sclerosing panencephalitis (although in this latter case it is not clear whether hypermutated genomic regions are essential for establishment of the "new" pathogen as a derivative of measles virus, or hypermutation is itself a byproduct of persistence in brain cells; see also Chapter 6). A limited number of amino acid substitutions at the E2 envelope glycoprotein and nsP3 protein of Venezuelan equine encephalitis virus are apparently able to trigger the epizootic emergence of equine encephalitis from an enzootic reservoir of the virus (Weaver, 1998; Wang et al, 1999).

In some cases, antigenic evolution may occur as a result of amino acid replacements at exposed antigenic sites, perhaps associated with vaccination or antiviral treatment. In some chronic carriers of HBV, surface antigen cannot be detected by routine serological assays. A large frequency of amino acid replacements within the major antigenic loop of the surface protein has been identified in such diagnostic-escape viruses (Weinberger et al, 2000). Some antigenic variants of HBV may be partially resistant to neutralization by antibodies induced by the current standard vaccine, and such variants may be gradually increasing in frequency in the human population (Zuckerman, 2000). Antigenic changes associated with HBV which contains mutations that diminish sensitivity to lamivudine (an antiviral nucleoside analog) have been also identified (Chen and Oon, 2000). Although possible connections between the two mutations have not been clarified yet, "hitchhiking" of genomes encoding an altered antigenic site in the virus variants selected by the antiviral agent could play a role in the dominance of these types of double mutant (see Fig. 6.7 in Chapter 6).

Most new human influenza pandemics (worldwide epidemics) have been associated with viruses which have acquired new genetic complement through segment reassortment [antigenic shifts due to replacement of one or more genes encoding the surface antigens, hemagglutinin (H) or neuraminidase (N): for example, the reassortment shift H1N1 → H2N2 originated the Asian influenza of 1957, and H2N2 → H3N2 originated the Hong Kong influenza of 1968 (Murphy and Webster, 1996; Webster, 1999)]. Canine parvovirus, with a single-stranded DNA genome, emerged as an important disease agent of dogs by mutation of a feline parvovirus in the 1970s, with no obvious environmental (or other) external influences apparently playing a role in this emergence (Parrish and Truyen, 1999). It is clear that in these two examples a contact between the parental virus and the potential new host was a necessary condition for the emergence. A reassortant influenza virus can originate only when one of the viruses in an animal reservoir (most commonly birds or swine) comes into replicative contact with another, related virus and coinfection of the same cell takes place (in humans or animals) to produce the reassortant. Similarly, a feline parvovirus must come into contact with dogs before an infection can be established. Viral traffic and new contacts are sine qua non conditions for viral emergence, but genetic variation of the pathogen often plays a role (Morse, 1993).

In many cases, environmental and demographic influences are quite obvious factors in emergence. Dams provide extensive water surfaces where larvae of insect vectors can proliferate. Expansion of Rift valley virus followed dam constructions in several African countries. Slave traffic from Africa to America resulted in the introduction of HTLV-I and yellow fever virus to the American continent. Whenever overpopulation and poor hygiene meet (such as in crowded prison camps of refugee camps during wars), infectious diseases are more likely to reemerge. It is difficult in these cases to distinguish the relative contributions of pathogen adaptive potential and environmental influences. There is obviously great indeterminacy regarding when and where new viral pathogens will emerge and whether, following the emergence, such outbreaks will remain localized (as in the case of those caused by Ebola virus) or will expand worldwide (as with the AIDS pandemic). It seems obvious that a number of public health and political precautions beforehand could help prepare for emerging infectious diseases (Fig. 8.7). There is little doubt that the evolutionary capacity of viral and cellular pathogens, in particular RNA viruses, will continue to pose important challenges, and that increased insight into their potential for rapid variation and evolution is needed.

Overview

There is now overwhelming evidence that RNA viruses are extremely error-prone in their replication; largely because their replication (unlike that of DNA-based lifeforms) does not involve proofreading or repair mechanisms to improve fidelity of genome copying. The extreme mutation rates of RNA genetic elements, together with their large population sizes, endows them with undeniable characteristics of quasispecies populations. Most progeny genomes of individual (clonal) RNA viruses generally differ from one another in having one or more mutations scattered randomly along their genomes; thus comprising a quasispecies "cloud" or "swarm" of related, but nonidentical variants. Darwinian selection acts upon these in a positive and negative manner to shape their evolution in any given environment. This evolutionary behavior of RNA viruses exhibits remarkable correspondence with the mathematical formalism of quasispecies theory as originated and elaborated by Eigen, Biebricher, Schuster and colleagues. Computer simulations, in vitro laboratory evolution of small RNA molecules, and the population behavior of living RNA viruses all confirm the validity and utility of quasispecies theory.

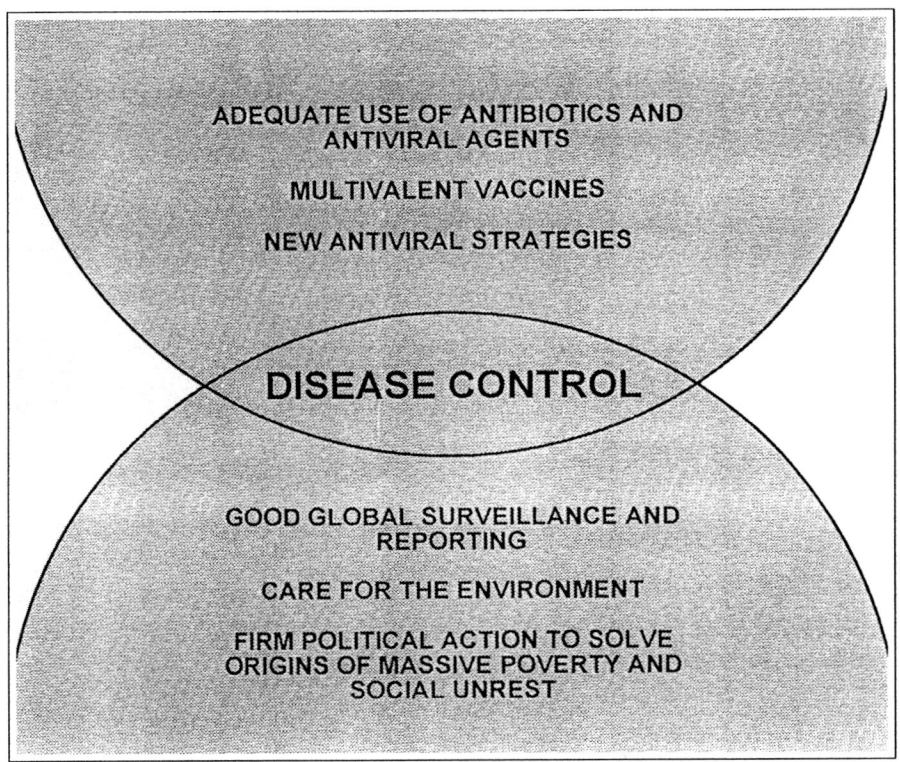

Fig. 8.7. Strategies for viral disease control. Additional information and relevant references are given in the text.

Quasispecies theory postulates that RNA replicases function very close to a mutational "error threshold". Near this threshold, RNA viruses attain maximal variability and the environment selects the most-fit progeny from a very large, very rich assortment of mutants. When this threshold is violated, there is a sudden phase transition which leads to melting of information and loss of viability in progeny genomes. Apparently, all RNA viruses have adopted this quasispecies strategy to maximize possibilities for rapid evolution, although rapid evolution is not necessitated if constant environments stabilize the persistence of highly fit "master sequences". In contrast, DNA viruses are less error-prone, and generally have adopted a more genetically-stable strategy in which they optimize function, then slowly "coevolve" along with their specific hosts over long time periods. Both strategies have been enormously successful, but it is the RNA strategy which most-often produces novelty (and sudden disease problems for hosts).

A key concept of quasispecies theory is the exploration of "sequence space" in searches for optimal fitness. Sequence space is incomprehensibly vast, having dimension of 4 to the vth power where v is the genome length, and its exploration would be impossible if it were not for its high connectivity (no position is farther away than the Hamming distance—the number of mutations which separate two sequences), and if searches were not "guided" by selection for fitness gains. Searches are guided by "upward movements" of mutant clouds along "fitness ridges" and "mountainous regions" of the "fitness landscape" of sequence space. This is analogous to occupation of "adaptive peaks" in the older theories of Sewall Wright. Nearly all of the total possible sequences of a 10 kb virus genome are "dead"—lacking information content.

Only a miniscule fraction of regions in sequence space could encode a living virus, and most subregions of these would be nonadaptive. However, despite their being only a tiny fraction of sequence space, there are countless viable regions (and adaptive subregions within these), and RNA virus quasispecies clouds are maximally-suited to search for the subregions confering high fitness. Viable, adaptive regions of sequence space are inevitably widely separated because they are such a tiny fraction of the whole, so the largest, most biologically significant evolutionary jumps are best achieved by mutants at the periphery of large quasispecies mutant clouds and by recombination/reassortment events. Such highly significant evolutionary jumps are, of course, considerably more rare than gradual, rather uneventful movements along fitness ridges and mountainous subregions of the adaptive landscape.

The vast size of sequence space and the large number of alternative fitness ridges to explore within any viable region preclude accurate predictions of future evolutionary pathways. This would be true even if the size of virus genomes were completely fixed and if the selective environment were completely constant. They are not, of course, and this confounds any hope for evolutionary predictability. Virus genomes expand and contract as chance and the environment dictate. The dynamic, ever-changing nature of our universe, and of our planet earth with its numerous competing life forms assures that the selective environment, can never be constant for long. Thus, adaptive landscapes change frequently, often drastically, so that even highly-selected, well-adapted master sequences and consensus sequences are temporary and provisional; here today—gone tomorrow—along with their vanished high-fitness peaks.

Among all of earth's life forms, the most genetically versatile and adaptable are the quasispecies mutant swarms of the RNA viruses. They were almost certainly among the first and almost certainly will be among the last, life forms on earth. From the beginnings, RNA genetic elements have explored sequence space rapidly and relentlessly; and they will do so till the end. Along the way, these evolutionary meanderings frequently generate formidable "new" (or more virulent) disease entities. This, too, will continue indefinitely, unpredictably and inexorably, so it is well for their most intelligent hosts to be at least modestly prepared. Likewise, it would be usefull to apply our increasing knowledge of quasispecies behavior toward improved drug and vaccine modalities, and toward combinatorial approaches to biotechnology and pharmaceutical developments including those for antiviral agents. They will be needed!

References

1. Anderson RM, May RM. Infectious diseases of humans: Dynamics and control. Oxford: Oxford University Press, 1991.
2. Arora R, Priano C, Jacobson AB et al. *Cis*-acting elements within an RNA coliphage genome: Fold as you please, but fold you must!! J Mol Biol 1996; 258:433-446.
3. Artenstein MS, Miller WS. Air sampling for respiratory disease agents in army recruits. Bacteriol Rev 1966; 30:571-575.
4. Ayala FJ, ed. Molecular Evolution. Sunderland: Sinauer Associates Inc., 1976.
5. Bak P. How Nature Works. The Science of Self-Organized Criticality. New York: Springer-Verlag, 1996.
6. Baranowski E, Sevilla N, Verdaguer N et al. Multiple virulence determinants of foot-and-mouth disease virus in cell culture. J Virol 1998; 72:6362-72.
7. Beaudry AA, Joyce GF. Directed evolution of an RNA enzyme. Science 1992; 257:635-641.
8. Bergstrom CT, McElhany P, Real LA. Tansmission bottlenecks as determinants of virulence in rapidly evolving pathogen. Proc Natl Acad Sci USA 1999; 96:5095-5100.
9. Biebricher CK, Eigen M, McCaskill JS. Template-directed and template-free RNA synthesis by Qβ replicase. J Mol Biol 1993; 231:175-179.
10. Biebricher CK, Gardiner WC. Molecular Evolution of RNA in vitro. Biophys Chem 1997; 66: 179-192.

11. Biebricher CK, Orgel LE. An RNA that multiplies indefinitely with DNA-dependent RNA polymerase: Selection from a random copolymer. Proc Natl Acad Sci USA 1973; 70:934-938.
12. Bol JF. Alfalfa mosaic virus and ilarviruses: involvement of coat protein in multiple steps of the replication cycle. J Gen Virol 1999; 80:1089-1102.
13. Bornscheuer UT. Directed evolution of enzymes. Angew Chem Engl Ed 1998; 37:3105-08.
14. Breaker RR, Joyce GF. Emergence of a replicating species from an in vitro RNA evolution reaction. Proc Natl Acad Sci USA 1994; 91:6093-97.
15. Breaker RR, Joyce GF. Self-incorporation of coenzymes by ribozymes. J Mol Evol 1995; 40:551-558.
16. Burgyan J, Salanki K, Dalmay T, Russo M. Expression of homologous and heterologous viral coat protein-encoding genes using recombinant DI RNA from cymbidium ringspot tombusvirus. Gene 1994; 138:159-163.
17. Burnet FM. The clonal selection theory of acquired immunity. Nashville, TN: Vanderbuilt University Press; 1959.
18. Cargill M, Altshuler D, Ireland J et al. Characterization of single-nucleotide polymorphisms in coding regions of human genes. Nature Genetics 1999; 22:231-238.
19. Chapman KB, Szostak JW. Isolation of a ribozyme with 5' → 5' ligase activity. Chem & Biol 1995; 2:325-333.
20. Chen WN, Oon CJ. Changes in the antigenicity of a hepatitis B virus mutant stemming from lamivudine therapy. Antimicrob Agnets and Chemother 2000; 44:1765.
21. Christians FC, Loeb LA. Novel human DNA alkyltransferases obtained by random substitution and genetic selection in bacteria. Proc Natl Acad Sci USA 1996; 93:6124-28.
22. Chu BCF, Kramer FR, Orgel LE. Synthesis of amplifiable reporter RNA for bioassays. Nucl Acids Res 1986; 14:5591-5603.
23. Clarke PH. Experiments on the evolution of bacteria with novel enzyme activities. Chem Scripta 1986; 26B:337-342.
24. Coen DM, Ramig RF. Viral genetics. In: Fields BN, Knipe DM, Howley PM et al, eds. Fields Virology. Philadelphia: Lippincott-Raven 1996; 113-151.
25. Collins PL, Mink MA, Stec DS. Rescue of synthetic analogs of respiratory syncytial virus genomic RNA and effect of truncations and mutation on the expression of a foreign reporter gene. Proc Natl Acad Sci USA 1991; 88:9663-67.
26. Convery MA, Rowsell S, Stonehouse NJ et al. Crystal structure of an RNA aptamer-protein complex at 2.8 Å resolution. Nature Struct Biol 1998; 5:133-139.
27. Cooper DN, Krawczak M. Human Gene Mutation. Oxford: Bios Scientific Publishers Limited, 1993.
28. Cowan GA, Pines D, Meltzer D, eds. Complexity. Metaphors, Models and Reality. Proceedings Vol XIX, Santa Fe Institute, Studies in the Sciences of Complexity. Reading MA: Addison-Wesley Publ. Co. 1994.
29. Crameri A, Raillard SA, Bermudez E et al. DNA shuffling of a family of genes from diverse species accelerates directed evolution. Nature 1998; 391:288-291.
30. Crick FHC. The origin of the genetic code. J Mol Biol 1968; 38:367-379,
31. Cuenoud B, Szostak JW. A DNA metalloenzyme with DNA ligase activity. Nature 1995; 375: 611-614.
32. Cupples CG, Miller JH. A set of *lacZ* mutations in *Escherichia coli* that allow rapid detection of each of the six base substitutions. Proc Natl Acad Sci USA 1989; 86:5445-5349.
33. Dahlberg AE. The functional role of ribosomal RNA in protein synthesis. Cell 1989; 57:525-529.
34. Domingo E. Biological significance of viral quasispecies. Viral Hepatitis Reviews 1996; 2:247-261.
35. Domingo E. RNA virus evolution and the control of viral disease. In: Jucker E, ed. Progress in Drug Research. Basel: Birkhauser Verlag 1989; 33:93-133.
36. Domingo E. RNA virus quasispecies as models of biological complexity. In: Dixon RA, Harrison MJ, Roossinck MJ, eds. 10th Anniversary Symposium Proceedings, The Samuel Roberts Noble Foundation, Plant Biology Division. Ardmore: The Samuel Roberts Noble Foundation; 1999: 79-90.
37. Domingo E. Variability and quasispecies. Current Opinion in Genetics and Development 1992; 1/2:61-63.

38. Domingo E. Viruses at the edge of adaptation. Virology 2000; 270:251-253.
39. Domingo E, Baranowski E, Ruiz-Jarabo CM et al. Quasispecies structure and persistence of RNA viruses. Emerging Infectious Diseases 1998; 4:521-527.
40. Domingo E, Escarmís C, Martínez MA et al. Foot-and-mouth disease virus populations are quasispecies. Current Topics in Microbiology and Immunology 1992; 176:33-47.
41. Domingo E, Escarmís C, Menéndez-Arias L et al. Viral quasispecies and fitness variations. In: Domingo E, Webster RG, Holland JJ, eds. Origin and Evolution of Viruses. London: Academic Press; 1999; 141-161.
42. Domingo E, Holland JJ, Biebricher C et al. Quasispecies: The concept and the word. In: Gibbs A, Calisher C, García-Arenal F, eds. Molecular Basis of Virus Evolution. Cambridge University Press, 1995:171-180.
43. Domingo E, Holland JJ. Complications of RNA heterogeneity for the engineering of virus vaccines and antiviral agents. In: Setlow JK, ed. Genetic Engineering, Principles and Methods. Plenum Press 1992; 14:13-32.
44. Domingo E, Holland JJ. RNA virus mutations and fitness for survival. Annu Rev Microbiol 1997; 51:151-178.
45. Domingo E, Martínez-Salas E, Sobrino F et al. The quasispecies (extremely heterogeneous) nature of viral RNA genome populations: Biological relevance: A review. Gene 1985; 40:1-8.
46. Domingo E, Menéndez-Arias L, Holland JJ. RNA virus fitness. Reviews in Medical Virology 1997; 7:87-96.
47. Doudna JA, Szostak JW. RNA-catalysed synthesis of complementary strand RNA. Nature 1989; 339:519-522.
48. Doudna JA, Usman N, Szostak JW. Ribozyme-catalyzed primer extension by trinucleotides—a model for the RNA catalyzed replication of RNA. Biochemistry 1993; 32:2111-15.
49. Drake JW. Rates of spontaneous mutation among RNA viruses. Proc Natl Acad Sci USA 1993; 90:4171-75.
50. Duarte EA, Novella IS, Weaver SC et al. RNA virus quasispecies: Significance for viral disease and epidemiology. Infectious Agents and Disease 1994; 3:201-214.
51. Edelman GM. The structure and function of antibodies. Sci Am 1970; 223(2):34-42.
52. Eigen M, Biebricher C. Sequence space and quasispecies distribution. In: Domingo E, Holland JJ, Ahlquist P, eds. RNA Genetics, vol. 3. Boca Raton: CRC Press Inc. 1988: 211-245.
53. Eigen M, Gardiner WC. Evolutionary molecular engineering based on RNA replication. Pure & Appl Chem 1984; 56:967-978.
54. Eigen M. Self-organization of matter and the evolution of biological macromolecules. Naturwissenschaften 1971; 58:465-523.
55. Ellinger T, Ehricht R, McCaskill JS. In vitro evolution of molecular cooperation in CATCH, a cooperatively coupled amplification system. Chem & Biol 1998; 5:729-741.
56. Ellington AD, Szostak JW. In vitro selection of RNA molecules that bind to specific ligands. Nature 1990; 346:818-822.
57. Encell LP, Coates MM, Loeb LA. Engineering human DNA alkyltransferases for gene therapy using random sequence mutagenesis. Cancer Res 1998; 58:1013-20.
58. Encell LP, Landis DM, Loeb LA. Improving enzymes for cancer gene therapy. Nature Biotechnol 1999; 17:143-147.
59. Evans DJ, McKeating J, Meredith JM et al. An engineered poliovirus chimera elicits broadly reactive HIV-1 neutralizing antibodies. Nature 1989; 339:385-388.
60. Ewald RW. Evolution of Infectious Disease. Oxford: Oxford University Press, 1994.
61. Fersht AR, Shi J-P, Wilkinson AJ et al. Analysis of enzyme structure and activity by protein engineering. Angew Chem Int Ed 1984; 23:467.
62. Fodor SPA, Read JL, Pirrung MC et al. Light-directed, spatially addressable parallel chemical synthesis. Science 1991; 251:767-773.
63. Forrest S. Genetic Algorithms: Principles of natural selection applied to computation. Science 1993; 261:872-878.
64. Garcia-Sastre A, Palese P. Genetic manipulation of negative-strand RNA virus genomes. Annu Rew Microbiol 1993; 47:765-790.

65. Garnet GP, Antia R. Population biology of virus-host interactions. In: Morse SS, ed. The Evolutionary Biology of Viruses. New York: Raven Press 1994: 51-73.
66. Gell-Mann M. The Quark and the Jaguar. Adventures in the Simple and the Complex. New York: Freeman, 1994.
67. Gerone PJ, Couch RB, Keefer GV et al. Assessment of experimental and natural viral aerosols. Bacteriol Rev 30:576-588.
68. Giver L, Gershenson A, Freskgard PO et al. Directed evolution of a thermostable esterase. Proc Natl Acad Sci USA 1998; 95:12809-813.
69. Goldenfeld N, Kadanoff LP. Simple lessons from complexity. Science 1999; 284:87-89.
70. Griffiths AD. Building an in vitro immune system—human antibodies without immunization from phage display libraries. Annales Biol Clin 1993; 51:554-554.
71. Gromeier M, Wimmer E, Gorbalenya AE. Genetics, pathogenesis and evolution of picornaviruses. In: Domingo E, Webster RG, Holland JJ, eds. Origin and Evolution of Viruses. San Diego: Academic Press 1999; 287-343.
72. Guatelli JC, Whitfield KM, Kwoh DY et al. Isothermal, in vitro amplification of nucleic acids by a multienzyme reaction modeled after retroviral replication. Proc Natl Acad Sci USA 1990; 87: 1874-78.
73. Guerrier-Takada C, Gardiner K, Marsh T et al. The RNA moiety of ribonuclease P is the catalytic subunit of the enzyme. Cell 1983; 35:849-857.
74. Hahn CS, Hahn YS, Braciale TJ et al. Infectious Sindbis virus transient expression vectors for studying antigen processing and presentation. Proc Natl Acad Sci USA 1992; 89:2679-83.
75. Hanski IA, Gilpin ME, eds. Metropopulation Biology. Ecology, Genetics and Evolution. San Diego: Academic Press, 1997.
76. Harayama S. Artificial evolution by DNA shuffling. Trends Biotechnol 1998; 16:76-82.
77. Harris H. Enzyme polymorphisms in man. Proc Roy Soc Lond B 1966; 164:298-310.
78. Holland JJ, Spindler K, Horodyski F et al. Rapid evolution of RNA genomes. Science 1982; 215: 1577-1585.
79. Holland JJ, Domingo E, de la Torre JC et al. Mutation frequencies at defined single codon sites in vesicular stomatitis virus and poliovirus can be increased only slightly by chemical mutagenesis. J Virol 1990; 64:3960-62.
80. Holland JJ, de la Torre JC, Steinhauer DA. RNA virus populations as quasispecies. Curr Top Microbiol Immunol 1992; 176:1-20.
81. Horwitz MSZ, Loeb LA. DNA sequences of random origin as probes of Escherichia colipromoter architecture. J Biol Chem 1988; 263:14724-731.
82. Illangasekare M, Sanchez G, Nickles T et al. Aminoacyl RNA synthesis catalyzed by an RNA, Science 1995; 267:643-647.
83. Iverson SA, Sastry L, Huse WD et al. A combinatorial system for cloning and expressing the catalytic antibody repertoire in *Escherichia coli*. Cold Spring Harbor Symp Quant Biol 1989; 54: 273-281.
84. Jerne NK. The generation of self tolerance and of antibody diversity. Eur J Immun 1971; 1:1-9.
85. Ji J, Hoffmann J-S, Loeb L. Mutagenicity and pausing of HIV reverse transcriptase during HIV plus strand DNA synthesis. Nucleic Acids Res 1994; 22:47-52.
86. Joyce GF. Directed molecular evolution. Sci Am 1992; 267(6):48-55.
87. Joyce GF. Nucleic acid enzymes -playing with a fuller deck. Proc Natl Acad Sci USA 1998; 95: 5845-47.
88. Kaslow RA, Evans AS. Epidemiologic concepts and methods. In: Evans AS, Kaslow RA, eds. Viral Infections of Humans: Epidemiology and Control. New York and London: Plenum Medical Book Co. 1997: 3-58.
89. Kaslow RA, Evans AS. Epidemiologic concepts and methods. In: Evans AS, Kaslow RA, eds. Viral Infections of Humans. Epidemiology and Control, 4 th edition. New York and London: Plenum Medical Book Company 1997; 3-58.
90. Kauffman SA. Applied molecular evolution. J Theor Biol 1992; 157:1-7.
91. Kilbourne DE. Host determination of viral evolution: a variable tautology. In: Morse SS, ed. The Evolutionary Biology of Viruses. New York: Raven Press 1994: 253-71.

92. Kohler G, Milstein C. Continuous cultures of fused cells secreting antibody of predefined specificity. Nature 1975; 256:495-497.
93. Korber B, Foley B, Hahn B et al, eds. Human Retroviruses and AIDS. A compilation and analysis of nucleic acid and amino acid sequences. Theoretical Biology and Biophysics Group T-10. Los Alamos NM: Los Alamos Natl Laboratory, 1997.
94. Koschel K, Brinckmann U, Hoyningen-Huene VV. Measles virus antisense sequences specifically cure cells persistently infected with measles virus. Virology 1995; 207:168-178.
95. Kruger K, Grabowski PJ, Zaug AJ et al. Self-splicing RNA: Autoexcision and autocyclization of the ribosomal RNA intervening sequence of Tetrahymena. Cell 1982; 31:147-157.
96. Kwoh DY, Davis GR, Whitfield KM et al. Transcription-based amplification system and detection of amplified immunodeficiency virus type 1 with a bead-based sandwich hybridization format. Proc Natl Acad Sci USA 1989; 86:1173-77.
97. Lamers SL, Sleasman JW, Goodenow MM. A model for alignment of env V1 and V2 hypervariable domains from human and simian immunodeficiency viruses. AIDS Res Hum Retroviruses 1996; 12:1169-78.
98. Leclerc F, Srinivasan J, Cedergren R. Predicting RNA structures: The model of the RNA element binding Rev meets the NMR structure. Folding and Design 1997; 2:141-147.
99. Lederberg J, Shope LE, Oaks SC, Jr., eds. Emerging Infections. Microbial threats to health in the United States. Washington, D.C.: National Academy Press, 1992.
100. Lee CH, Gilbertson DL, Novella IS et al. Negative effects of chemical mutagenesis on the adaptive behavior of vesicular stomatitis virus. J Virol 1997; 71:3636-40.
101. Lerner RA, Tramontaur A. Catalytic Antibodies. Sci Am 1988; 258(3):42-50.
102. Levin BR, Lipsitch M, Bonhoeffer S. Population biology, evolution, and infectious disease: Convergence and synthesis. Science 1999; 283:806-809.
103. Levis R, Huang H, Schlesinger S. Engineered defective interfering RNAs of Sindbis virus express bacterial chloramphenicol acetyl transferase in avian cells. Proc Natl Acad Sci USA 1987; 84: 4811-15.
104. Lewontin RC, Hubby JL. A molecular approach to the study of genic heterogeneity in natural populations. II. Amount of variation and degree of heterozygosity in natural populations of Drosophila pseudoobscura. Genetics 1966; 54:595-609.
105. Lewontin RC. The Genetic Basis of Evolutionary Change. New York: Columbia University Press, 1974.
106. Li RX, Dowd V, Stewart DJ et al. Design, synthesis, and application of a protein A mimetic. Nature Biotechnol 1998; 16:190-195.
107. Lipshutz RJ, Morris D, Chee M et al. Using oligonucleotide probe arrays to access genetic diversity. Biotechniques 1995; 19:442-447.
108. Lizardi PM, Kramer FR. Exponential amplification of nucleic acids—new diagnostics using DNA polymerases and RNA replicases. Trends Biotechnol 1991; 9:53-58.
109. Loeb LA. Unnatural nucleotide sequences in biopharmaceutics. Advances in Pharmacology 1996; 35:321-347.
110. Loeb LA, Essigmann JM, Kazazi F et al. Lethal mutagenesis of HIV with mutagenic nucleoside analogs. Proc Natl Acad Sci USA 1999; 96:1492-97.
111. Lorsch JR, Szostak JW. In vitro evolution of polynucleotide kinase ribozymes. FASEB J 1995; 9: A1422-22.
112. Luytjes W, Krystal M, Enami M, Parvin J, Palese P. Amplification, expression, packaging of a foreign gene by influenza virus. Cell 1989; 59:1107-13.
113. Mahy BWJ. Human viral infections: An expanding frontier. Antiviral Res 1997; 22:47-52.
114. Martínez MA, Hernández J, Piccone ME et al. Two mechanisms of antigenic diversification of foot-and-mouth disease virus. Virology 1991; 184:695-706.
115. Martínez MA, Dopazo J, Hernández J et al. Evolution of the capsid protein genes of foot-and-mouth disease virus: Antigenic variation without accumulation of amino acid subtitutions over six decades. J Virol 1992; 66:3557-65.
116. Martínez MA, Verdaguer N, Mateu MG et al. Evolution subverting essentiality: Dispensability of the cell attachment Arg-Gly-Asp motif in multiply passaged foot-and-mouth disease virus. Proc Natl Acad Sci USA 1997; 94:6798-6802.

117. May RM. Ecology and evolution of host-virus associations. In: Morse SS, ed. Emerging Viruses. Oxford: Oxford University Press 1993: 58-68.
118. May RM. The coevolutionary dynamics of viruses. In: Gibbs A, Calisher CH, García-Arenal F, eds. Molecular Basis of Virus Evolution. Cambridge: Cambridge University Press 1995: 192-212.
119. Maynard Smith J, Burian R, Kauffman S et al. Develpmental constraints and evolution. The Quart Rev of Biol 1985; 60:265-287.
120. Meyerhans A, Vartanian J-P. The fidelity of cellular and viral polymerases and its manipulation for hypermutagenesis. In: Domingo E, Webster RG, Holland JJ, eds. Origin and Evolution of Viruses. San Diego: Academic Press 1999; 87-114.
121. Milstein C. Monoclonal antibodies. Sci Am 1980; 243(4):66-74.
122. Morse SS, ed. Emerging Viruses. Oxford: Oxford University Press, 1993.
123. Moya A, García-Arenal F. Population genetics of viruses: An introduction. In: Gibbs A, Calisher CH, García-Arenal F, eds. Molecular Basis of Virus Evolution. Cambridge: Cambridge University Press 1995:213-23.
124. Mullis K, Faloona F, Scharf S et al. Specific enzymatic amplification of DNA in vitro: the polymerase chain reaction. Cold Spring Harbor Symp Quant Biol 1986; 51:263-273.
125. Munir KM, French DC, Loeb LA. Thymidine kinase mutants obtained by random sequence selection. Proc Natl Acad Sci USA 1993; 90:4012-16.
126. Munishkin AV, Voronin LA, Ugarov VI et al. Efficient templates for Qβ replicase are formed by recombination from heterologous sequences. J Mol Biol 1991; 221:463-472.
127. Murphy FA, Nathanson N. The emergence of new viral diseases: an overview. Seminars in Virology 1994; 5:87-102.
128. Murphy FA. New, emerging and reemerging infectious diseases. Adv Virus Res 1994; 43:1-52.
129. Murphy RB, Webster RG. Orthomyxoviruses. In: Fields BN, Knipe DM, Howley PM et al, eds. Fields Virology. Philadelphia: Lippincott-Raven; 1996: 1397-1445.
130. Nei M. Molecular Evolutionary Genetics. New York: Columbia University Press, 1987.
131. Nicolis G, Prigogine I. Self-oragnization in hon-equilibrium systems. New York: Wiley, 1999.
132. Niewlandt P, Decker D, Gold L. In vitro selection of RNA ligands to substances. Biochemistry 1995; 34:5651-59.
133. Noller HF. Transfer RNA ribosomal RNA interactions and peptidyl transferase. FASEB J 1993; 7: 87-89.
134. Novella IS, Domingo E, Holland JJ. Rapid viral quasispecies evolution: Implications for vaccine and drug strategies. Molecular Medicine Today 1995; 1:248-253.
135. Parmley SF, Smith GP. Antibody-selectable filamentous fd phage vectors-affinity purification of target genes. Gene 1988; 73:305-318.
136. Parrish CR, Truyen U. Parvovirus variation and evolution. In: Domingo E, Webster RG, Holland JJ, eds. Origin and Evolution of Viruses. San Diego: Academic Press; 1999:421-439.
137. Pathak VK, Temin HM. 5-Azacytidine and RNA secondary structure increase the retrovirus mutation rate. J Virol 1992; 66:3093-3100.
138. Pattnaik AK, Ball LA, Le Grone AW et al. Infectious defective interfering particle of VSV from transcripts of a cDNA clone. Cell 1992; 69:1011-20.
139. Pawlovsky J-M, Germanidis G, Neumann AU et al. Interferon resistance of hepatitis C virus genotype 1b: Relationship to nonstructural 5A gene quasispecies mutations. J Virol 1998; 72:2795-2805.
140. Penny MA, Thomas SJ, Douglas NW et al. Env gene sequences of primary HIV type 1 isolates of subtypes B, C, D and F obtained from the World Health Organization Network for HIV isolation and characterization. AIDS Res Hum Retroviruses 1996; 12:741-747.
141. Piccirilli JA, Krauch T, Moroney SE et al. Enzymatic incorporation of a new base pair into DNA and RNA extends the genetic alphabet. Nature 1990; 343:33-37.
142. Racaniello VR, Baltimore D. Cloned poliovirus complementary DNA is infectious in mammalian cells. Science 1981; 214:916-919.
143. Radecke F, Spielhofer P, Schneider H et al. Rescue of measles virus from cloned DNA. EMBO J 1995; 14:5773-84.
144. Rost B, Sander C. Bridging the protein sequence-structure gap by structure predictions. Annu Rev Biophuys Biomol Struct 1996; 25:113-136.

145. Roy SK, Harris SG. Antisense epidermal growth factor oligodeoxynucleotides inhibit follicle-stimulating hormone-induced in vitro DNA and progesterone synthesis in hamster preantral follicles. Mol Endocrinol 1994; 8:1175-1181
146. Ruiz-Jarabo CM, Sevilla N, Dávila M et al. Antigenic properties and population stability of a foot-and-mouth disease virus with an altered Arg-Gly-Asp receptor-recognition motif. J Gen Virol 1999; 80:1899-909.
147. Ruiz-Jarabo CM, Arias A, Baranowski E et al. Memory in viral quasispecies. J Virol 2000; 74:3543-47.
148. Saiki RK, Gelfand DH, Stoffel S et al. Primer-directed enzymatic amplification of DNA with a thermostable DNA polymerase. Science 1988; 239:487-491.
149. Sambrook J, Fritsch EF, Maniatis T. Molecular cloning: A laboratory manual (Second edition). Cold Spring Harbor, NY: Cold Spring Harbor Press; 1989.
150. Santoro SW, Joyce GF. Mechanism and utility of an RNA cleaving DNA enzyme. Biochemistry 1998; 37:13330-342.
151. Schnell MJ, Mebatsion T, Conzelmann K-K. Infectious rabies viruses from cloned cDNA. EMBO J 1994; 13:4195-4203.
152. Schober A, Walter NG, Tangen U et al. Multichannel PCR and serial transfer machine as future tool in evolutionary biotechnology. BioTechniques 1995; 18:652-660.
153. Schultz PG. Catalytic antibodies. Angew Chem Int Ed 1989; 28:1283-95.
154. Schultz PG, Lerner RA. Antibody catalysis of difficult chemical transformations. Acc Chem Res 1993; 26:391-395.
155. Schuster P. Evolutionary biotechnology-Theory, facts and perspectives. Acta Biotechnol 1996; 1: 3-17.
156. Schuster P, Hogeweg P, von Gabain A et al. Molecular Evolution and Biotechnology. Trends, vol. 5. Science, Research, Development. Published by the European Commission, Directorate General XII, Belgium, 1998.
157. Schuster P, Stadler PF, Renner A. RNA structures and folding. From conventional to new issues in structure predictions. Curr Op in Struct Biol 1997; 7:229-235.
158. Scott JK, Smith GP. Searching for peptide ligands with an epitopic library. Science 1990; 249: 386-390.
159. Sheld WM, Armstrong D, Hughes JM, eds. Emerging Infections 1. Washington, D.C.: ASM Press, 1998.
160. Shope RE, Evans AS. Assessing geographic and transport factors, and recognition of new viruses. In: Morse SS, ed. Emerging Viruses. Oxford: Oxford University Press 1993: 109-19.
161. Simon HA. The Sciences of the Artificial. Cambridge: MIT Press, 1996.
162. Smith GP, Scott JK. Libraries of peptides and proteins displayed on filamentous phage. Meth Enzymol 1993; 217:228-257.
163. Spielhofer P, Bächi T, Fehr T et al. Chimeric measles virus with a foreign envelope. J Virol 1998; 72:2150-59.
164. Spiess EB. Genes in populations. New York: John Wiley and Sons, 1977.
165. Stark BC, Kole R, Bowman EJ et al. Ribonuclease P: An enzyme with an essential RNA component. Proc Natl Acad Sci 1978; 75:3717-21.
166. Swetina J, Schuster P. Self-replication with error—a model for polynucleotide replication. Biophys Chem 1982; 16:329-345.
167. Szostak JW. Ribozymes evolution ex vivo. Nature 1993; 361:119-120.
168. Taniguchi T, Palmieri M, Weissmann C. Qβ DNA-containing hybrid plasmids giving rise to Qβ phage formation in the bacterial host. Nature 1978; 274:223-228.
169. Tuerk C, Gold C. Systematic evolution of ligands by exponential enrichment: RNA ligands to bacteriophage T4 DNA polymerase. Science 1990; 249:505-510.
170. Tyagi S, Kramer FR. Molecular beacons-probes that fluoresce upon hybridization. Nature Biotechnol 1996; 14:303-308.
171. Verdaguer N, Mateu MG, Andreu D et al. Structure of the major antigenic loop of foot-and-mouth disease virus complexed with a neutralizing antibody: Direct involvement of the Arg-Gly-Asp motif in the interaction. EMBO J 1995; 14:1690-96.

172. Verdaguer N, Mateu MG, Bravo J et al. Induced pocket to accomodate the cell attachment Arg-Gly-Asp motif in a neutralizing antibody against foot-and-mounth disease virus. J Mol Biol 1996; 256:364-376.
173. Verdaguer N, Sevilla N, Valero ML et al. A similar pattern of interaction for different antibodies with a major antigenic site of foot-and-mouth disease virus: Implications for intratypic antigenic variation. J Virol 1998; 72:739-748.
174. Volkenstein MV. Physical Approaches to Biological Evolution. Berlin: Springer-Verlag, 1994.
175. Wagner EGH, Simons RW. Antisense RNA control in bacteria, phages, and plasmids. Annu Rev Microbiol 1994; 48:713-742.
176. Wang E, Barrera R, Bashell J et al. Genetic and phenotypic changes accompanying the emergence of epizootic subtype IC Venezuelan equine encephalitis viruses from an enzootic subtype ID progenitor. J Virol 1999; 73:4266-71.
177. Weaver SC. Recurrent emergence of Venezuelan equine encephalomyelitis. In: Sheld WM, Hughes J, eds. Emerging infections. I. Washington DC: American Society for Microbiology Press; 1998: 27-42.
178. Webster RG. Antigenic variation in influenza viruses. In: Domingo E, Webster RG, Holland JJ, eds. Origin and Evolution of Viruses. San Diego: Academic Press 1999:377-390.
179. Weinberger KM, Bauer T, Böhm S et al. High genetic variability of the group-specific a-determinant of hepatitis B virus surface antigen (HBsAg) and the corresponding fragment of the viral polymerase in chronic virus carriers lacking dtectable HBsAg in serum. J Gen Virol 2000; 81:1165-74.
180. Wilson C, Szostak JW. In-vitro evolution of a self-alkylating ribozyme. Nature 1995; 374:777-782.
181. Wu Y, Zhang DY, Kramer FR. Amplifiable messenger RNA. Proc Natl Acad Sci USA 1992; 89: 11769-773.
182. Zaug AJ, Grabowski PJ, Cech TR. Autocatalytic cyclization of an excised intervening sequence RNA is a cleavage-ligation reaction. Nature 1983; 301:578-583.
183. Zhang BL, Cech TR. Peptidyl-transferase ribozymes—trans reactions, structural characterization and ribosomal RNA like features. Chem & Biol 1998; 5:539-553.
184. Zuckerman AJ. Effect of hepatitis B virus mutants on efficacy of vaccination. The Lancet 2000; 355:1382-84.

Index

A

Antibiotic resistance 133, 134
Antigenic variation 107-110, 143

B

Bacteria 63, 64, 66, 93, 104, 120, 132-134, 156, 157
Biological Information 76
Bottleneck 94, 100, 109, 120-123, 125, 141

C

Catalysis 5, 35, 37-39
Central dogma 4, 11
Compensatory mutation 34, 42
Competition 2, 15, 19, 20, 25, 38, 39, 51, 71-73, 78, 82, 85, 100, 101, 119, 120, 124, 128, 131, 132, 141, 147, 148
Competitive exclusion 128
Complementation 15, 20, 43, 44
Complexity 65, 76, 98, 102, 105, 107, 123, 129, 131, 132, 142, 146-148, 158, 160
Copying fidelity 38, 41, 43

D

DNA polymerase 5, 22, 37-41, 61
DNA viruses 93, 108, 129, 130, 151, 160, 163

E

Emergence of new RNA virus species 2
Equilibrium population 56
Error rate 20, 39, 44, 60-62, 72, 158
Error threshold 56, 61, 158, 160, 163
Evolutionary biotechnology 148, 149
Expression 3, 4, 7, 8, 11-13, 22, 23, 25, 28, 30, 33, 34, 37, 43, 50, 65

F

Fitness 20, 44, 50, 52, 57, 59-62, 71, 74-78, 85-89, 93, 95, 96, 102, 103, 105, 108, 119-128, 134, 136, 141, 145, 149-152, 156, 159, 163, 164

G

Genetic code 4, 30, 31
Genome 2-4, 7-13, 15-24, 28, 30, 32, 33, 38, 39, 42-44, 55, 57, 59-61, 63-65, 74, 76
Genotype 2, 3, 56, 60, 61, 63, 65, 66, 72, 74, 76, 77
Growth rate 48-51, 53, 55, 67, 69, 71

H

Heredity 2
Hypermutation 83, 97, 131, 161

M

Molecular clock 92
Molecular recognition 33
Mutation 2, 15, 18, 20, 24, 25, 30, 33, 34, 38, 42-44, 49, 51-57, 59, 61, 63, 64, 66, 71, 72, 74-78, 82-89, 91-98, 100, 101, 103, 109, 119-124, 126-134, 141-147, 150-152, 156-163

N

Negative selection 85, 93, 95, 103, 110, 127, 141
Negative strand 11, 16, 19

O

Origin of species 2

P

Phenotype 2, 3, 9, 33, 51-54, 60, 63, 65, 74, 77
Phenotypic variant 147
Polymerization 38-40
Polymorphism 99, 144, 145
Polyprotein 13, 15, 17
Population landscape 59, 60
Positive selection 95, 103, 108, 127, 134
Positive strand 11-13, 15-17, 19, 42

Index

Q

Quasispecies 9, 10, 16, 19, 22, 24, 25, 34, 44, 50, 56, 57, 59, 60, 67, 74-76, 82, 85, 93, 95, 98-100, 102-110, 120-123, 125-131, 141, 142, 144-147, 158, 162-164

R

Rate of evolution 89, 100, 158
Reassortment 82, 83, 87, 89, 91, 161, 162, 164
Recombinant DNA 150, 157
Recombination 22, 39, 55, 61, 76, 77, 82, 83, 86-89, 94, 111, 120, 121, 129-131, 134, 151, 157, 160, 161, 164
Red Queen 128
Replicase 4, 12, 14, 15, 19, 33, 39-41, 43, 56, 63-72, 74, 76-78, 83, 88, 101, 127, 152, 153, 163
Reverse transcriptase 4, 11, 21, 22, 38-41
Ribose 4, 5
RNA replication 9, 13, 32, 39, 41, 61, 64-67, 69, 101, 102, 152, 153
RNA structure 14, 31-33
RNA synthesis 5, 15, 20, 40, 43, 63, 67, 72

S

Selection 2, 4, 19, 20, 23-25, 33, 34, 38, 39, 43, 44, 50-54, 56, 57, 59, 61, 62, 64, 66, 70-74, 76, 78, 121, 125, 127, 131-134, 141, 146-148, 152, 155-157, 159, 162, 163
Sequence space 57, 59, 60, 63, 75, 76, 78
Shannon entropy 147

T

Template 4, 11, 13, 16, 17, 19, 21-23, 35, 37, 38, 40-43, 60, 65-70, 74, 76-78, 80, 82-88, 97, 101, 103, 154-156
Transcriptase 4, 11, 18, 21, 22, 38-41
Transmission 94, 104, 127, 134, 141, 142, 164

V

Viral pathogens 141, 160
  emergence 143, 144, 158, 160, 162
Viral rececptor 71
Viral receptor 7
Virus 2-5, 7-25, 29, 31-34, 37-44, 47, 49, 55, 56, 60-63, 65, 71, 76, 78, 80